GENESIS
AND
THE
BIG BANG

———■———

GENESIS

AND

THE

BIG BANG

**The Discovery of Harmony Between
Modern Science and the Bible**

GERALD L.
SCHROEDER, Ph.D.

BANTAM BOOKS
NEW YORK • TORONTO • LONDON • SYDNEY • AUCKLAND

GENESIS AND THE BIG BANG
A Bantam Book
Bantam hardcover edition published October 1990
Bantam trade paperback edition / January 1992

ISBN 0-553-35413-2

Published simultaneously in the United States and Canada

Bantam Books are published by Bantam Books, a division of Bantam
Doubleday Dell Publishing Group, Inc. Its trademark, consisting of
the words "Bantam Books" and the portrayal of a rooster, is Registered
in U.S. Patent and Trademark Office and in other countries. Marca
Registrada. Bantam Books, 666 Fifth Avenue, New York, New York 10103.

Dedication

■■■■■■■■■■■■■■■■■■■■■■■

To my children,
whose probing questions led to the writing of this book.

Contents

■■■■■■■■■■■■■■■■■■■■■■

Acknowledgments

....

Genesis and the Big Bang has evolved over the past decade from a few notes jotted in response to questions from my children to the form it has today. In this ten-year period, I have never changed the title because my goal has remained the same: Finding and exposing the unity that binds the biblical Genesis and cosmological theory. It is, in fact, the same unity that binds all mankind and all the universe.

I have had much help in this effort, not the least of which has come from my wife, Barbara, whose efforts have kept our home life on course. The training I received during my undergraduate and graduate studies at M.I.T. gave me the tools needed for scientific research. The subtleties of Scripture and the subtleties of science have much in common. My aspiration to engage in research in one or both of these fields was fostered early on by my parents, who had a reputation among their closest of associates as being natural philosophers. Later on, Professor Robley Evans, under whom I completed my doctorate research, and Rabbi Chaim Brovender and the late Rabbi Herman Pollack, under whom I did my first serious biblical research, were guides in learning to sift facts from what is often a plethora of less than relevant information. Adelaide and Seymour Kahn contributed their resources along the way.

In the formal preparation of *Genesis and the Big Bang*, three individuals were decisive and essential contributors: Helen Rees, who believed in and guided the project from the earliest stage; Marc Jaffe, who read, commented on, and reread the manuscript as it evolved; and Michelle Rapkin of Bantam Books, who edited the manuscript and whose suggestions at times gave me the impression of a gentle intellectual breeze moving aside a curtain and allowing an unobstructed Sun to illuminate the text. To all who helped me I offer my sincere thanks and hope that their efforts were not wasted.

Jerusalem, Israel

Who was it that said, "If you can only afford one newspaper, read the opposition's"?

Introduction

∎∎∎∎∎∎∎∎∎∎∎∎∎∎∎∎∎∎∎∎∎∎∎

Study astronomy and physics if you desire to comprehend the relation between the world and God's management of it.

—Maimonides
The Guide for the Perplexed

U ntil 1961, my image of a deer was a blend of the healthy, full-coated stag in the Bronx Zoo and Walt Disney's Bambi. During that year, on a clear, cool autumn morning in the Nevada desert, this image was indelibly changed. I was sitting on the roof of a Department of Defense (DOD) pickup truck that had been parked since the previous night atop Rainier Mesa, located in the midst of the Atomic Energy Commission's Nevada Nuclear Testing Site—for the first time in three years, the United States had detonated an atomic bomb.

I was part of any liberal's dream come true. The goal of my work was to develop a method for locating the epicenters of underground nuclear explosions. This technology would be essential if the United States and the Soviet Union were ever to reach an accord for nuclear disarmament. Mutual disarmament without a workable treaty might rapidly degenerate into unilateral disarmament, and this could be more destructive to peace than an all-out arms race. One of the stumbling blocks for such a treaty is the system of verification; there must be a method to prove that no signee to the treaty has reneged and conducted a clandestine weapon test. Such a test would most likely occur underground. The earth, in theory, can absorb so much of the energy and radiation released by the blast that the test might go undetected by alien observers. My colleagues and I reasoned, correctly we were to learn, that we could identify the epicenter of such an explosion by detecting subtle changes in the soil's structure caused by the

seismic wave of the weapon's blast. This would be effective even
if all the bomb-produced radiation was contained underground
within the huge glass ball that forms as the heat of the blast melts
the adjacent rock.

Because I was working toward the establishment of an enforce-
able treaty for disarmament, I could morally justify being part of
the excitement—and there was to be much of that—of the nuclear
scene. In this profession, we rarely referred to the weapon as a
bomb or the blast as an explosion. The weapon was a "device"
and the blast an "event."

United flight 711 arrived at Las Vegas's McCarran International
Airport late at night. The glow from the strip motels gave the
place an aura of fantasy—more like an amusement park than a
city. That aura of fantasy was to remain even as I was lining up
for admission to my sixth atomic blast. But this time it was my
first exposure, and, in fact, the first for many of us involved.

I picked up the keys to my Avis Chevy and packed the four
sacks of commercial cement I had brought with me. That ship-
ment had cost the taxpayers $400 in overweight baggage, but I
needed that particular cement because I knew its radioactive content. I had measured it a few days before in the "iron room" of
the M.I.T. Radioactivity Center. The four sides, floor, and ceiling
of the iron room were made of 5-inch-thick slabs of iron cut from
the shielding of a mothballed battleship. This shielding absorbs so
much of the natural radiation background of both cosmic and earthly
origins that even the small amount of radiation associated with the
radium contained in a healthy human's skeletal frame can be quantitatively measured. The sacks of cement I was bringing produced
a radioactive signal of about four times that of a human's. Knowing
this, I could calculate the effect it would have on the data I was
here to gather at the weapon test.

You have to travel 50 miles northwest on U.S. 95 from Las
Vegas to the entrance gate of the test site. In those days it was a
two-lane road that followed the contour of the land. There were
no bridges. At each culvert crossing, you dipped down the bank
of the wadi, temporarily disappearing from the sight of any on-
coming traffic, and then a fraction of a second later, with the
springs of your car bottomed out from the force of the sharp ascent,
you shot back into view. There was no speed limit for most of

the way. It was such a routine trip that we would drive it at flat-out speed. Only on the stretch passing the half-dozen buildings of Indian Springs and Cactus Springs was a speed limit imposed and that was 60 mph. The sheriff in each of these towns drove a pickup. Two rifles were always conspicuous in the rack at the window behind the driver. One was high powered with a telescopic sight; the other was a shotgun. In the heat of the desert, tempers can run high. Violence was not uncommon.

After passing through the security check at the entrance gate to the test site, there were another thirty miles of two-lane road leading to Rainier Mesa. Civilian cars were not permitted on the mesa during the test. I left my rented Chevy at the base camp and continued in the DOD pickup truck that had been assigned to me.

For this planned weapon test, a tunnel had been dug a few hundred meters into the mesa, a third of the way up the face. The bomb was placed at the end of the tunnel, and part of the tunnel had been backfilled to keep the force of the blast from shooting straight out the hole. I needed gas sampling tubes set 8 meters into the top of the mesa just above where the bomb was to explode. That required a drill rig that could cut through the outcrop of welded-tuff rock that overlay the epicenter of the planned event. We had only a week's notice prior to the first detonation and so the pace was feverish. It took a fifth of Jack Daniels to convince the drilling foreman to work overtime and sink the holes for me. In 24 hours I was in business.

Weapon testing takes place in the very early morning, usually just after sunrise. This timing gives the personnel involved in the work a full sunny day to track the effects of the event. By mid-afternoon of the day prior to the test, all nonessential personnel had been cleared off the top of the mesa. Four scientists, of which I was one, would need access to ground zero immediately after the event. Along with three rad-safe personnel, we were to sleep on the mesa that night. Concern for any potential danger did not enter my mind. The physicists in charge of estimating the energy release of the event claimed that they would be willing to stand at the epicenter and merely bend their knees to absorb the shock.

It was a wonderfully clear, windless day and the stillness of the scene was as intense as the afternoon heat. Any scrub growth there was had the gray-green color of tough desert vegetation covered with dust—rain was still a few months off. Far below us, the deserted dirt access road cut a solitary line across the miles of Frenchman's and Yucca flats and then angled steeply up the face of the mesa.

The night sky of a desert has a clarity that city dwellers never know. In the dry desert air there is so little reflected light that the sky is either complete black or a pure white diamond star. The Milky Way cuts a clear path across the sphere of stars and the double star in the Big Dipper can be recognized with no added magnification. When God took Abraham out of his tent and told him that the multitude of his progeny would be as the numbers of stars, this is what he must have seen. "And [Abraham] believed in God, and He counted it to him as righteousness" (Gen. 15:6). For an elderly, childless man to believe that he would have such a number of progeny must have indeed taken complete faith.

The cool of a desert morning is a fleeting pleasure. All during the night, the exposed rock and soil radiate heat through the clear, dry air out to the blackness of space. It is as if the Earth were involved in a futile attempt to repay the Sun for its warming gift of the previous day. Because desert soil has no significant amount of moisture, which stores the Sun's heat in wetter climes, the temperature of the ground rapidly drops at night and in turn cools the overlying air. The result: a much-appreciated nighttime chill that gives way to the heat of the Sun within moments after sunrise.

By the time the Sun rose on this special autumn day, I had finished my bagged breakfast. I had slept out on the open flatbed of the DOD pickup that had been assigned to me. Now, alongside two other DOD pickup trucks, I was sitting on the roof of the cab looking out over a mile and a half of scrub toward the epicenter. Five seconds before detonation, high-speed, tripod-mounted movie cameras started to whir. These would create a visual record of ground motion. Four flashbulbs fired, marking the moment of detonation on film. At first it seemed as if that was the end, for aside from the noise of the cameras, the desert was silent.

Then it arrived. I imagine the mixed sensations of wonder and fear that we felt are akin to those experienced on seeing the ocean recede prior to the arrival of a tidal wave. My truck and those beside me started to rock as if they were at sea. A blast 2000 meters away and hundreds of meters below us had turned the mesa into a bowl of shaking Jell-O. The face of the mesa fell away to the alluvial flat below, a millennium of erosion in an instant, a sort of time warp in the jargon of relativistic physics. The access road, the only exit we knew, was gone. We held on to the sides of the trucks.

A deer, dusty gray and scrawny with short broken antlers, bolted from the scrub onto the flat rock outcrop we had claimed as our clearing and stopped for a moment; a young fawn followed. I read terror in the animals' unblinking stares as they raced past us and disappeared into the brush.

Something had not followed plan. A few grams of uranium were to have been converted into a sudden burst of exquisitely hot energy, the E of Einstein's famous $E = mc^2$. The amount of energy release had been calculated to produce an explosion easily contained within the mass of the mesa. But apparently there had been a slipup. The blast had caused the collapse of the front of the mesa and we were now in the midst of a man-made earthquake. None of this had been predicted, or if it had, we had not been informed of it. Neutron capture by the uranium atoms held in the center of the bomb is the key to the chain reaction that leads to a nuclear explosion. The more neutrons that are captured and absorbed by the uranium, the more intense the reaction and the greater the energy release. To put it mildly, from what we were experiencing, it seemed to me that predicting the efficiency of neutron capture was still an inexact science.

At first, none of us spoke, and then we all spoke at the same time.

"Those guys said they would be willing to stand at the epicenter and bend their knees. It would have bent a lot more than their knees."

"I need measurements close to time zero. Can we drive in?"

"Nothing doing until we verify that the area is rad safe."

"But we were guaranteed the blast wouldn't vent."

"Procedures require a rad-safe survey."

Two rad-safe men, now in white coveralls, headed toward the epicenter along a double-rutted lane, Geiger counters at the ready. Twenty minutes later they came back, clicking. Their Geiger counters, covered with radioactive dust, were now useless. The meters were swamped with their own radioactivity. The third rad-safe man and one from the first pair went back toward the epicenter. Ten minutes later they returned. The results were the same—it was now clear that a radioactive cloud or front was moving toward us although there was no perceptible breeze. Not only had the force of the blast obliterated the face of the mesa, but it also had fissured the mesa's top. These fissures were allowing radioactive gases to vent to the atmosphere.

One Geiger counter remained uncontaminated, and we put it into a plastic lunch bag. This way, if the bag got contaminated we could change it and still keep the meter in service. At this point, we had had enough of the epicenter and just wanted to get out of there. Someone in our group found a map of the mesa in the glove compartment of his truck. It showed two thin lines leading off the back of the mesa. They were trails, not roads. Geiger counter in the lead, we bounced along in a caravan retreating from the newly released radiation.

We like to think that man is in control and that we can control the bomb. Perhaps ... but at that moment the bomb was controlling us.

On the drive to Las Vegas, I told my technician not to discuss the events of the day with his wife, but I knew there was little chance of his keeping this experience to himself.

One of the pleasures of the Vegas motels is their poolside decks. Considering the pace of work for the last five days, I voted myself the right to a poolside drink. Two-thirds through my second whiskey sour, I was paged.

"This is the major here."

"Yes sir. Quite a day, wasn't it?"

"Please don't mention the events of today to anyone until we meet to brief you on what is to be classified."

"Yes sir."

"Enjoy your evening."

With that brief conversation, I had my first personal immersion into the realpolitik of the weapons race.

The evening news and the next day's papers reported the test: As planned there was no release of radioactivity, no danger to the populace. It was to be a decade before declassification revealed that all the tests vented radiation into the atmosphere.

*   *   *

The bomb is powerful. And as mankind has learned on several unfortunate occasions, its functioning is not always within our technological control. It does not necessarily produce the expected results. Unfortunately, nuclear weapons do not stand alone in their promise of an undelivered product. Since the industrial revolution we have turned in vain to science in a search for cornucopias to fulfill our needs. We long for a great technological leap forward to free us from want and to bring peace on Earth. The *Titanic* was to remove the peril of travel at sea. But this unsinkable boat was sunk on its maiden voyage after hitting an iceberg that had drifted a bit off its proper course. The *Electra* was to be the uncrashable plane. It fell apart in clear skies; a stress point had been overlooked. It was only a few short decades ago that elimination of hunger and human drudgery seemed at last to be within our reach. Microbiology and nuclear power were the respective agents for these tasks. Electricity would be too cheap to bother with metering. All we needed to do was to achieve the control that many of us working in nuclear physics had taken for granted as attainable in nuclear reactors. The events of Windscale, Three Mile Island, and Chernobyl have made us rethink the validity of this means to the end of drudgery. Microbial single-celled proteins (SCP) might have fed us all. They double their mass in just fractions of an hour. There seemed to be almost no limit to the quantities of these nutritious foods that could be produced. But again there was a catch. The very molecules required by the SCP for their rapid reproduction, the abundant nucleic acids, so adversely affect the metabolic processes of both lower animals and humans that the consumption of SCP in large quantities is not possible. We thought that nuclear weapons and gas, because of their devastating effects, would force peace on us. But peace has still eluded our grasp. Again, the failed promise of an undelivered cornucopia.

Although technology has failed to bring salvation from material want, the overall record of technology is not one of failure. The multitude of its successes have greatly improved the quality of life for the relatively small fraction of humanity fortunate enough to share in its benefits.

Unfortunately, the profusion of the material benefits that we experience in the twentieth-century Western world has not brought in its wake the holistic tranquility we seek. The widespread abuse of drugs and alcohol and the all-too-common cynicism and depression in today's societies stem from the failure of our newly found material bounty to satisfy our multifaceted needs. It may have the sound of a triviality, but in fact we have rediscovered the deep reality that "not by bread alone does man live" (Deut. 8:3). Science and technology have diminished the physical burdens in our lives; however, the enigmatic relation between the brain and the mind, between, for example, the physical gratification of a desire and the resulting sensation of happiness, indicates that there is an aspect to life that exceeds that which is intellectually tangible. It has been a long, slow process, but we are finally realizing that within each of us there is a spiritual as well as a material aspect. Not one of us is an exception. If a human is to function as a harmonious unit, if the sought-after tranquility is to be found, then both the spirit and the body must be nourished.

Finding a source for our spiritual guidance is not as simple as finding a supplier for our material needs. As technology progresses, the new usually displaces the old and society benefits from the advance. The technological advances of science have been so rapid and materially impressive that, in almost all aspects of life, "new" has come to mean "better." The impact of "new as better" is nowhere more blatant than in the withering of cultural traditions.

The scientific revolution brought the proof that mankind was physically not at the center of the universe. Having replaced the old geocentric theory with the more modest understanding of our physical position in the universe, the concept parallel to this was, of course, that we were also not spiritually central to the universe. What a burden to have lifted from our shoulders! Not being central, not being the target creation of a Creator, if there was one, meant freedom from obligations toward that creation.

Unfortunately, wisdom is not bound by the flow of the calendar. What is new intellectually may have displaced traditionally accepted norms, but this newness may bring with it neither truth nor benefits. Having fallen victim to the misconception that the new is inherently an advance over the old, most of us have politely placed biblical tradition in the pigeonhole of irrelevance. To reach the conclusion that this tradition is an atavistic, superstition-tinged set of concepts is to trivialize a source of information that has borne relevance for 3400 years of human existence. It is a tradition that has at its base the common quest of all decent people: Peace on Earth, good will to all. "You shall love your neighbor as yourself" (Lev. 19:18) is the great principle of the Torah. Judging from our mixed achievements in this quest for a world free of want, exploitation, and oppression, there remains much to learn, spiritually as well as materially.

In the Talmud we are given a clue for finding the spiritual guidance, the soul food as it were, that we are lacking. There we are told that Hillel the Elder, in the first century B.C.E.,[1] was asked by a skeptic to teach the entire Torah, the guide that had directed Hillel in his life's journey, and to do this while he stood on one foot. Hillel fulfilled the request. "What is hateful to you," he said, "do not to your friend. This is the entire Torah. The rest is commentary. Go forth and learn."[2]

In our quest to go forth and discover the tie that links our heritage as human beings to the stuff of the universe, we will study both the wisdom of this tradition and the knowledge of today's science and find that the functioning of the universe has a relevance for us all. Contained in its history is a path that has traversed many troubled waters and ultimately led to our being. For in the most literal sense, we and all life are composed of star dust. From an understanding of our cosmic path, we may decipher whether there is, in fact, a purpose for our existence and perhaps even learn what that purpose may be.

We will find that the synthesis of scientific discoveries and biblical tradition reveals a universe that has evolved from a state of chaos to one that today shines forth as a cosmos, as a system that is composed of many parts and yet functions in harmony and order. It is a unity that we might well emulate in our personal journey through the space-time continuum of our lives.

Two very different types of researchers toil at understanding our cosmic history. One delves into the secrets of the universe through physics and cosmology. The other relies on interpretations of the

Bible. In spite of their common goal, they use such different sources of information that they often appear to be mutually antagonistic. I strongly believe that the antagonism experienced by these two groups is unnecessary. Rather, it is the result of a sort of professional provincialism, a myopic view of knowledge. An understanding of both physics and biblical tradition shows that the opening chapters of the book of Genesis and the findings of modern cosmology corroborate rather than dispute each other.

I make this statement well aware that skeptics from both the scientific and the biblical schools are looking over my shoulders. They have a shared design: Each wants to leave the topic totally unconvinced of the validity of the other's teachings. I doubt that I can convince an avowed secularist that, beyond providing cultural insights, the Bible is also a valid source of cosmological insight. The fundamentalist, on the other hand, may have little use for the claims of cosmology as an aid to understanding the Bible. If this book can broaden the perspective by which each views the other's knowledge, then I will have accomplished my goal.

CHAPTER 1

The Search

■■■■■■■■■■■■■■■■■■■■■■

Know what to answer a skeptic.
—Talmud[3]

For the last few years, I have had a continuing debate with my son Joshua. By the age of eight, Josh had already completed five years of biblical study. He had been taught to relate to the Bible in its most literal sense, and so for him, and for many of his teachers, the age of the universe is exactly the age derived from the generations as they are listed in the Bible. This comes to approximately 57 centuries. For them, the cosmological estimate of the age of the universe, some 15 billion years, is a preposterous fiction. According to a literal reading of the Bible, six days is stated as the time between "the beginning" and the appearance of man—and six days it is. As we shall learn, the forebears of this same biblical tradition were well aware that a theology devoid of knowledge of the physical universe was (and is) a contradiction in terms. The implications of general relativity and Doppler shifts in light are an essential part in understanding the opening chapters of Genesis. Although the insights provided by modern science have yet to be assimilated by many contemporary teachers of Genesis, we shall find that they are a help rather than a hindrance toward the rapprochement of the secular and nonsecular.

The discrepancy between the 5700-year biblical age of the universe and the 15-billion-year scientific estimate might not have caused the family debate that led me to write *Genesis and the Big Bang* had there not been one, almost extraneous, circumstance in our lives. That circumstance was a nearby archaeological dig.

For three years we lived in Zikhron Ya'aqov, a village on the Carmel range overlooking Israel's coastal plain. The plain is bounded on the west by the Mediterranean. A few kilometers inland from the shore, the massive limestone outcrop of the Carmel range rises abruptly to form the plain's eastern boundary. It is a wonderfully productive area. This is especially true of the region between Zikhron Ya'aqov and Haifa. Here rows of avocado and orange trees, heavy with foliage and fruit, border the blue-green waters filling acres of fishponds. These ponds share space with white-specked fields of cotton. The productivity of the area is probably not new. Archaeological evidence indicates that this land has provided its material abundance to man and his predecessors for tens of thousands of years.

A few kilometers north of our village, among scrub growth halfway up the steep face of Mt. Carmel, the yawning black openings of several particularly large caves appear as sentinels overseeing the coast. On two or three occasions, my family and I have scrambled up Mt. Carmel to explore the archaeological digs within the caves. It is these digs that are the source of my continuing debate with Josh.

Here artifacts and fossils of that prehistoric animal, Neanderthal man, have been found. For the theologian steeped in a superficial reading of Genesis, these fossils may be Neanderthal, but they have nothing to do with man!

For the paleontologist, the age of the fossils also presents a problem, but of a different sort. The bones of Neanderthal man are too old to date using carbon-14 methods. So much of the original skeletal carbon-14 has decayed (one-half the original content decays radioactively each 5,600 years) that the small amount of carbon-14 remaining cannot be reliably measured. The bones' location within geologically known strata gives them an estimated age of 60,000 years.[4] For Josh and for many other people, Jew and Christian alike, a 60,000-year archaeological history is at least 54,000 years in error. It is totally at odds with a literal understanding of the opening chapters of the book of Genesis.

But this discord between archaeology and theology is neither necessary nor valid. As I have studied the details of biblical and scientific texts, I have reached what was for me an enlightening and unexpected conclusion: The biblical narrative and the scientific account of our genesis are two mutually compatible descriptions of the same, single, and identical reality. My goal in this book is to explain that compatibility to expert and layperson alike.

Simply stated, I will discuss here the first week of Genesis. To the literalist, these first six days can be a problem. Fossils place the appearance of beings with humanlike features at well over the 60,000-year age of the Neanderthal fossils of the Carmel caves. Although dating the time of Peking man's existence has the same types of problems as those encountered when dating the remains of Neanderthal man, the archaeological estimate is that these humanlike animals lived some 300,000 years ago. Their home, near Cho-k'ou-tien, China, was in a group of limestone caves quite similar to those found in the Carmel range. *Homo erectus*, the genus and species to which Peking man belonged, made its appearance 1.5 to 2 million years ago. Because of this antiquity, only a few nearly complete *Homo erectus* skeletons have been found. One of these, located in what is now an arid region of Kenya, is the Lake Turkana Boy. It is estimated that the lad lived his twelve years of life 1.6 million years ago. Looking at this fossil, it is clear that, except for the shortened forehead, these early hominids had features so similar to modern humans that they could go unnoticed if they walked on a busy city street.

But does mankind's history really stretch back a million years, or even several tens of thousands of years? A strictly literal view of the Bible would say no. To the literalist, there was no prehistoric man. Adam was the first man and he was formed some 5700 years ago from the dust of the Earth (Gen. 2:7).

Part of the contradiction between archaeological evidence and biblical tradition is semantic, originating in the use of the word *man*. Archaeologists use the term to describe prehistoric animals that had some, but not necessarily all, of the features of modern man. Biblically speaking, the term *man* is applicable only to Adam and his progeny. It is, however, no secret that the divergent views of science and tradition are based on differences far more fundamental than the definition of words. In fact, most biblical commentators accept that the meaning of words of the Bible, especially as in the creation story, must be understood in accordance with their context. The terms for *water*, *darkness*, *wind*, *heavens*, and *earth* as used in the first ten verses of Genesis have meanings that are quite different when used later on in the Bible. It is the calendar of the early universe that raises a discrepancy far more difficult to explain than the variances in terminology.

Entire herds of fossilized rhinoceroses, their calves nestled between their legs in nursing position, have been dated by the potassium-argon content of the volcanic ash that buried them, as having lived 10 million years ago. Dinosaurs have been firmly established, at least according to paleontologists' data, to have roamed the Earth for 100 million years, becoming extinct some 65 million years ago. The simplest forms of life, the early prokaryotic (nucleus-free) bacteria and blue-green algae, make their first fossil appearance in earliest sedimentary rocks. These are dated as being 3 to 4 billion years old. The oldest rocks are only some 600 million years older. These are the original igneous rocks.

The length of this fossil record certainly seems to contradict the relatively brief period of creation and genesis presented by the Bible. And yet the fossil record covers only a fraction of the age that current cosmological observations ascribe to our universe. Where do the preponderance of paleontological and astronomical data, indicating an Earth some 4.5 billion years old and a universe that reaches back 10 to 20 billion years, fit into traditional biblical scholarship? At first it would appear that these findings put paleontology hopelessly at odds with the Bible. For the total cosmic evolution from the creation (the "In the beginning . . ." of Genesis 1:1) to the appearance of Adam gets only a bit more than five biblical days.

As an affirmation of faith, the literalist can explain the paleontological finds as having been placed in rocks at creation by the Creator. They might be there to satisfy man's need to rationalize the nature of the world, or to test man's belief in the biblical narrative. This argument, while impossible to disprove, is the weakest of reeds (Isa. 36:6) in a world full of explosive and convincing discovery.

Some theologians argue that the methods of paleontological dating are flawed. These methods depend on measurements of radioactive decay (see Table 1). They claim that the rates of decay today are not the same as in prehistoric times. If this were true, the ages of fossils or rocks could not be estimated from current measurements of radioactivity.

TABLE 1. HOW THE RADIOACTIVE DECAY OF CARBON IS USED TO DATE A FOSSIL

	Atoms		Atoms		Atoms
	0000	first	00	second	0
Carbon-14	0000	5600 years	00	5600 years	0
	0000		00		0
	0000	first	0000	second	0000
Carbon-12	0000	5600 years	0000	5600 years	0000
	0000		0000		0000

Two of the types of carbon found in all living things are referred to as carbon-12 and carbon-14. Both carbon-12 and carbon-14 act the same way as far as the chemical reactions of life are concerned, but carbon-14 is radioactive and carbon-12 is not. During every 5600 years, half of the carbon-14 atoms decay. Therefore, from the time that a plant or an animal dies, the ratio between the amounts of carbon-14 and carbon-12 is constantly decreasing. During the first 5600 years following death, the ratio decreases by half because half the carbon-14 decays in this period. During the next 5600 years, this ratio again decreases by half so that at the end of this period the carbon-14 to carbon-12 ratio will be one-fourth of the ratio that existed at the time of death. This continual decrease proceeds as long as any carbon-14 remains. By measuring the ratio of carbon-14 to carbon-12 in a fossil, it is possible to estimate the amount of time that has passed since death. Two assumptions are necessary to use radioactivity as a means of dating: The rates of decay must always have been the same as they are now and the ratio of carbon-14 to carbon-12 (or whichever other radioactive elements are being used) at the moment of death must be known. We assume that the initial ratio of carbon-14 to carbon-12 in plants and animals has always been similar to what it is today.

Again, it is impossible to disprove the idea that patterns of radioactive decay have changed during the past few thousand years. But the very concept of a fickleness in nature is contrary to all modern evidence. Our experience with the laws of nature, including those that govern radioactivity, is that they are unchanging. Imagine the bedlam of our lives if we were forced to test the consistency of gravity each time we put a glass on a table or the rate of passage of time in our Newtonian system each time we had an appointment to keep. The constancy of nature's laws is an integral part of life as we experience it. In fact, we base our entire life pattern on the assumption that the laws of nature are predictable.

But we need not reconcile biblical tradition and the findings of science with weak reeds. It is possible to maintain faith in the validity of the Bible and also to accept that the cave on Mt. Carmel *does* have 60,000 years of history layered in the deposits on its floor and that this cave was a dwelling place of the prehistoric creature we have labeled Neanderthal man.

Both scientific and religious traditions are systems of thought that seek the truth. Religion bases its search for truth within knowledge believed to have been attained by revelation. Science seeks the same truth by evaluating interactions observed in our physical world. Often, it is not realized that gaining an accurate understanding of the Bible is an endeavor that can be as demanding as the research of science. The biblical text is tersely cogent and yet "it speaks in the language of man."[5] Included in the seemingly simple accounts of mankind's development and encounters with the Creator are concepts that, we see by hindsight, were to guide the Western world for over 3000 years and to speak to civilizations that stretch from a tribe of just-freed slaves trekking through an unsown desert (Jer. 2:2) to modern man circling the Earth in a space capsule.

To the questions that science has posed to tradition there are answers, but they emerge only from a serious probe of the science and an understanding of the tradition that goes well beyond the literal biblical text.

The realization that biblical scholarship and the natural sciences are closely related is not new—but it is still important. "Study astronomy and physics if you desire to comprehend the relation between the world and God's management of it." This deep insight into sources of knowledge was written in the year 1190 by Moses Maimonides (*The Guide for the Perplexed*). Thousands of years earlier, the author of the book of Ecclesiastes warned, "If the iron is blunt and he does not sharpen its edge, then he must exert extra effort, but," he continued, "wisdom increases his skill" (Eccles. 10:10). Although it is true that, as the Psalmist wrote, "The heavens tell of the glory of God and the firmament tells the works of His hands" (Ps. 19:2), it is also a reality that understanding the story that these heavens tell is no small task.

To reach our goal of understanding the interplay, the intimate connection, between the Divine and the natural, we will need more than a love for the Bible; we also need a knowledge of the natural sciences. Maimonides compared those who understand the blending of the natural and the Divine to royal subjects being within the throne room of their exalted king, while those seeking an understanding of the Bible but lacking an understanding of the natural sciences were subjects groping in vain for the outermost gate of their king's palace.[6]

Without an understanding of nature, the Earth appears to be the center of the universe. Man appears to be able to solve all material problems and learn all details. His works become his gods. Galileo's probing of space with the telescope dispelled the first notion. Heisenberg's uncertainty principle dispelled the second. Unfortunately, we are at times still shackled to the third notion of our works being our gods.

If we are to approach the "throne room" of which Maimonides wrote, we will have to understand at least part of the physics and chemistry of inert and living forms of matter. We must deal with exotic-sounding concepts such as *entropy* and *general relativity*. To be certain that we consider only established principles in the natural sciences, I have imposed limits on the sources of information we use. These are publications of leading scientists in their fields, including works by physicists and cosmologists such as Albert Einstein, Steven Weinberg, Stephen Hawking, Edwin Taylor, John Archibald Wheeler, and Alan Guth; geophysicists such as A. G. W. Cameron, Frank Press, and Raymond Siever; and biologists and molecular chemists such as George Wald and Francis Crick.

When we seek an understanding of the biblical text, we are confronted with the abbreviated manner in which episodes are recorded in the Bible. This succinct form has made interpretive explanations of the text an integral part of biblical study. In Jeremiah (23:29) we read: "Behold, My words are as fire, says the Lord, and as a hammer that shatters the rock." Based on the parallel context of these two phrases, the Talmud teaches that "just as a hammer breaks a rock into many pieces, so can a single biblical passage have many meanings."[7] The result is that there is a plethora of opinions on each biblical passage.

In this book, I deal almost exclusively with information contained in the Five Books of Moses, that is the Pentateuch, or in Hebrew, the Torah. The validity and import of these five books, and really of the entire Old Testament, is shared by Jew and Christian alike. Because of this shared heritage, it has been possible here to rely on commentaries that are in essence nonsectarian in that they embrace beliefs held in common by the Judeo-Christian tradition.

Only a few biblical commentators have withstood time's test. Four are accepted by Jew and Christian alike as guiding lights in the interpretation of the book of Genesis. It is on these four that I rely. They are Onkelos (ca. C.E. 150), Rashi (Solomon ben Isaac, C.E. 1040–1105, France), Maimonides (Moses ben Maimon, C.E. 1135–1204, Spain, Egypt; also known as Rambam), and Nahmanides (Moses ben Nahman, C.E. 1194–1270, Spain, Israel; also known as Ramban). Because their commentaries were written long before the advent of modern physics, we avoid the folly of using interpretations of tradition that may have been biased by modern scientific discoveries.

Onkelos translated the Five Books of Moses into Aramaic. Aramaic was the common language of the Middle East during the centuries that preceded and followed the start of the common era. It was the language routinely spoken by Jews of the period in which the debates in the Talmud were recorded. This included the time of Jesus. Aramaic, a semitic language, is sufficiently similar to Hebrew to allow it to serve as a linguistic cross-reference. By translating the Hebrew into Aramaic, Onkelos provided definitions to words, the meanings of which had become obscure. This source is still in such standard use that it is referred to as "the translation."

Rashi's trailblazing commentary on the Bible made available to all students interpretations of biblical words and phrases that

hitherto had been accessible only to students learning with leading scholars.

Maimonides, in his *Guide for the Perplexed* (1190), discussed
aspects of the Bible that, as the title suggests, perplexed students
and laypersons. The sections of the *Guide* relevant to our study
deal with aspects of Genesis and include the time shortly after
the initial act of creation.

Nahmanides, teaching in the century after Rashi and Maimonides, had the benefit of their scholarship coupled with his personal training in tradition and mysticism. He believed that a full
understanding of the origins of the universe was contained in the
Bible as received by Moses. The information was written either
explicitly or by hints taken in some instances from the very form
of the written text.

> Hear counsel and receive instruction in order that you will be-
> come wise
>
> (Prov.19:20).

In recent times an unusual orientation has arisen among many
who are eager to evaluate the relevance of the Bible to daily life.
There is the misconception that an understanding of the text
comes to one as a natural heritage, as if we are bequeathed it
genetically along with the instinct to breathe or the ability to
reason. The most simplistic meanings are assumed to be adequate
and study of commentaries thereon is considered superfluous.
Now this presents an interesting inconsistency in how we evaluate information.

The paleontologist shows that the brontosaurus bones he has
examined have a scientifically confirmed age of 80 million years
and, an adherent to the theory of evolution, interprets these and
other fossils to demonstrate a theory of evolution. But there is no
dynamic pro-Darwinian evidence in the fossil record. Neither the
fossils nor the variety of life that surrounds us provides any proof
of one species changing into another, or a development of complex life forms from earlier, more simple forms. The incomplete
data are analyzed and interpreted, and we now feel secure, based
on these interpretations, that life did indeed start with a few

simple morphologies and develop over millennia to the seemingly infinite variety of today's living forms.

Similarly, the cosmologist interprets spectra of starlight to prove that the universe is 10 to 20 billion years old and in a state of rapid expansion. But a literal interpretation of the star-filled sky, whether viewed by eye or by telescope, is that the Sun and Moon and planets move and all other celestial bodies occupy fixed, unchanging positions. Perhaps three or four times per thousand years one of these tiny specks of light we call stars gets brighter and then fades to oblivion. But for the most part, the heavens are quiescent. They give the *literal* appearance of a steady-state system, not one in expansion as *interpretations* of cosmological data imply.

On the basis of a literal "reading" of the physical world, we might discard the claims of scientists to a deeper understanding of our universe. But such literal interpretations of the heavens or of biology are never used by scientists. Interpretation is an integral, indeed essential, part of scientific inquiry.

This flow between the apparent and the actual holds for biblical study as well. A literal reading of the Bible reveals only a part of the wealth of information held within the text. Every major theological treatise that has withstood the test of time promulgates the need for an understanding of the biblical text in a way that is compatible with *both* literal and interpretative meanings.

Laws of physics, especially those describing the flow of time and matter, are often difficult to conceptualize. They are, however, amenable to proof. What might be called the laws of divine science, that is, the validity of biblical interpretations, are considerably less open to proof. In fact, they are usually taken (or rejected) on faith.

If we are not to construct arguments of gossamer, we must tackle some of the more rigorous aspects of theology. These theological concepts are even more difficult to integrate into our thought processes than those of physics and cosmology. We all have an emotional stake in religion. Be it pro or con, our psyches will resist changes in our perceptions of the Bible's meaning. The most basic theological concept that we require as a background to our discussions is the realization that the Bible includes a range of knowledge, only part of which is immediately obvious from the actual text. Perception of this information requires an interpretive understanding of the text.

An analogy of such interpretation might be a layman reading a text written in scientific notation. For example, the statement 10^3 could be read as a typographical error for 103. To the trained scientist, it is obvious that 10^3 is read as 1000. The information held within the biblical text, but beyond the literal reading of the words, was as obvious to those versed in biblical interpretation as the fact that, to anyone who knows scientific notation, 10^3 means 1000 and not 103.

In his summary of the Torah (Deut. 31), Moses in effect calls out for interpretation of the text by referring to the entire Torah as a poem. Because the text is clearly not written in the form of a poem, this reference appears to relate not to form but to "the essence." The meanings of poems go well beyond the literal text and include subtleties held within the words and even the form. This is what Maimonides taught in regard to Proverbs 25:11, "A word fitly spoken is like apples of gold in a filigree vessel of silver"—the vessel (the literal meaning of the text) is beautiful and valuable, but the golden apples held within the vessel (the inner meanings of the text) are even more beautiful and valuable.

Nahmanides taught that the subtleties found in the Torah go even beyond those of a poem, reaching to the very shapes of the letters. In the "Introduction" to his *Commentary on Genesis*, he stated that everything that was transmitted to Moses was written in the Torah explicitly or by implication in words, in the numerical value of the letters, as bent or crooked letters, or their crownlets. This was implied in the verse from the Song of Songs (1:4): "The King has brought me into His chambers." This verse informs us that in the Bible, the king, God, has given mankind the knowledge of God's way in the universe, God's chambers, as it were. We need only know how to interpret the Bible.

Because of the importance accorded not only to the content of the Bible, but also to its form, to this day the addition or removal of a single letter, or even the changing of the shape of a letter, invalidates an entire Torah scroll. For a text containing 304,805 letters, this was and is an extraordinary demand of precision.

Torah scrolls are still written only by specially trained *sofreim*, or "writers of the Book." The parchment and even the ink must be specially prepared for use in the scroll.

With the establishment of the State of Israel, scrolls of the Five Books of Moses have been brought together from communities separated in time since the exile of Jews from Jerusalem and in space by the thousands of miles between the Arabian peninsula and Western Europe. Perhaps the most extreme of the reunions was the return of the Yemenite community to Israel. The Jewish community of Yemen had a residence in Arabia since the destruction of the Temple in 586 B.C.E.

The texts and forms of the scrolls from the Arabian peninsula differed in only one aspect from those saved from the Holocaust of Europe. The cover of an Arabian scroll is rigid and as such it holds the scroll in a vertical position when opened to be read or studied. The cover of a European scroll is soft cloth and because it offers no support, the scrolls when opened are placed horizontally on a table. But as for the actual content and form, they are one and the same. If indeed the Torah as revealed at Sinai contained concealed information, the faithfulness by which the *sofreim* transmitted the text across the generations has ensured that this information is still potentially available.

Unfortunately we no longer have the skills to lay open all the hints within the Bible. We have, however, the teachings of the sages who tapped this information. The topics of these teachings range from the cleansing of a clay pot to the creation of the universe. As we shall see, combined they present an ancient account of our cosmic roots, which are remarkably similar to those presented by current cosmology and paleontology.

In a manner that parallels the unfolding of deeper meanings within the Bible, the advances achieved by scientific research during the last 50 years have brought major changes in our understanding of our universe, its age, origin, and development. Well into the twentieth century, both astronomers and physicists believed the age of the Earth to be measured in millions, not billions, of years. Oceanographers gave the origin of the ocean waters as condensation from a primeval cloud that surrounded a once molten Earth. Biologists gave the source of life as the result of random associations of molecules such as ammonia and methane, that eventually through trial and error grew to be amino acids, the structures of life, and then life itself. Spacial dimensions and the passage of time were considered absolute, fixed, and mathematically true. As the core sciences of physics, chemistry, and mathematics made their way into what were once primarily descriptive fields of astronomy, geology, and biology, the age of the Sun was found to be measured in billions and not millions of years; the amount of water able to be contained in a cloud surrounding a molten Earth was found to be a small fraction of the water now present in the world's oceans; the time for random associations even of optimal, chemically advanced molecules (such as amino acids) to form the simplest bacterium was found to be possible only in times that equaled or exceeded the entire age of the universe— not merely the time the Earth, or even the solar system, has existed. But most extraordinary of all, the rate of time's passage and the spacial dimensions of objects were found to be not at all fixed but instead to vary in a grand manner that depends on the relation of the observer and the observed.

These new scientific insights might have been uncomfortable for traditionalists of science. They represented radical departures from the ideas prevalent at the time. In sharp contrast to this, for scholars who understand biblical tradition, each advance has increased the compatibility of science and theology.

For over 3000 years following the revelation at Mt. Sinai, the Western world had based its traditional concept of the origin of the world on the opening chapters of Genesis. This posed no intellectual problem because there was no challenge from a credible outside source that staked an equal claim to fundamental truths. It was a time when the family tree showing a few generations could satisfy a desire to understand one's position within the world. Remaining questions were answered by a religion that came neatly packaged in prayer books and a pastor's sermon. Then came the scientific enlightenment, and physics and cosmology claimed to provide the fundamental truths concerning our existence. Cosmologists measured the age of the universe to be 10 to 20 billion years and not the approximately 5700 years calculated from the dates given in the Bible. Paleontologists discovered evidence of a gradual evolution in the forms of life. This evolution extended over billions of years, not a few days. Man became merely the latest addition to the tree of life. Finally, the crushing blow: All life was found to have a single and extraordinarily complex genetic code—clear indication of man's evolution from bacteria.

Since the rise of scientific thinking and discovery, particularly in the late nineteenth and early twentieth centuries with the giant intellectual leaps taken by Einstein, Bohr, Planck, and others, the issues surrounding the very early universe have become crucial to our understanding of mankind's place on Earth, the nature of the cosmos, and the meaning of Genesis.

For the average layperson, Jew or Christian, there must necessarily be a conflict between science and biblical tradition. Is the biblical story of Creation the ultimate metaphor? Is it to be taken as literal truth? How can we reconcile the observable facts of paleontology and the laboratory proofs of the equations of Einstein with the very essence of Judeo-Christian faith—the biblical story of the first six days?

Have these seemingly unequivocal scientific revelations sounded the death knell for religious tradition as we have known it? Are the worlds of science and religion mutually exclusive?

The truth is just the opposite. The newly won knowledge of the universe is in fact the fertile ground for tradition's flowering.

It is essential to bear in mind that science has not provided explanations for the two principal starting points in our lives: the start of our universe and the start of life itself. When we try to describe the conditions at that crucial interface between total nothingness and the start of our universe, we are confronted with a point of space having infinitely high density and infinitely small dimensions. In the language of physics, such a point is called a singularity. Singularities cannot be handled mathematically in the dimensions we experience: the length, width, and height of things and the passage of time. Changing to imaginary dimensions of time allows the math to be handled but does nothing to remove the fact that an untenable singularity existed in real time at the Big Bang.

We might bypass this problem of a beginning by hypothesizing that there was no beginning, that the universe has been here forever. But even that does not work. This would require an infinite number of cycles in which the universe expanded, then contracted and reexploded to start a new Big Bang cycle. But, as we shall learn, this is thermodynamically not possible. There could not have been an infinite number of cycles if any of the material world were to remain. Because we are very much here in our tangible forms, along with all the other material stuff of the world, there must have been a beginning sometime in the finite past. But we cannot understand how, scientifically.

The answers provided by science for life's origins are no more satisfying than those provided for the universe's origins. Since the monumental "Conference on Macro-Evolution" was held in Chicago in 1980, there has been a total reevaluation of life's origins and development. In regard to the Darwinian theory of evolution, the world-famous paleontologist of the American Museum of Natural History, Dr. Niles Eldridge, unequivocally declared, "The pattern that we were told to find for the last one hundred and twenty years *does not exist.*"[8] There is now overwhelmingly strong evidence, both statistical and paleontological, that life could not have been started on Earth by a series of random chemical reactions. Today's best mathematical estimates state that there simply was not enough time for random reactions to get life going as fast as the fossil record shows that it did. The reactions were either directed by some, as of yet unknown, physical force or a metaphysical guide, or life arrived here from elsewhere. But the "elsewhere" answer merely pushes the start of life into an even more unlikely time constraint.

For decades, many scientists have presented the misconception that there are rational explanations for the origins of the universe, life, and mankind. The shortcomings of the popular theories were merely swept under the rug to avoid confusing the issues. The knowledge that scientists do *not* have these explanations has now been coupled by the awareness of the fossil record's failure to confirm Darwin's (or any other) theory of the gradual evolution of life. The demonstration of these misconceptions has brought many scientists and laypersons to an uncomfortable realization: The problems of our origins, problems that most of us would have preferred to consider solved by experts who should know the answers, in fact have *not been solved* and are not about to be solved, at least not by the purely scientific methods used to date.

During my three decades as a scientist active in applied nuclear physics and oceanography, and a 25-year immersion in the study of biblical tradition, I have found that there are answers to the questions posed by my son Joshua. What was for me the most exciting discovery in this search is that the duration and events of the billions of years that, according to cosmologists, have followed the Big Bang and those events of the first six days of Genesis are in fact one and the same. *They are identical realities that have been described in vastly different terms.*

The scientific and theological sources that led me to this realization are not the intellectual reserve of an elite in either discipline. Rather they are a heritage readily available to all who seek them. Using these sources, we can complete what might otherwise be a perplexing search for our cosmic roots. The result can be, for skeptic or believer, Jew or Christian, a fresh understanding of the key events in the life of our universe and in one's personal genesis as well.

Notes

1. To maintain the nonsectarian nature of this book, the labeling of dates herein uses the terminology B.C.E. for "before the common era" (in place of B.C.) and C.E. for during the "common era" (in place of A.D.).
2. *Babylonian Talmud*, Section Sabbath 31a.
3. *Babylonian Talmud*, Section Avoth 2:19.
4. Paleontological and archaeological dates used in this book represent averages taken from a variety of sources, including textbooks, scientific journals, and personal research in museums. A list of the published sources appears in the Bibliography.
5. *Babylonian Talmud*, Section Sanhedrin 64b.
6. Maimonides, Mishnah Torah 4:10, and *The Guide for the Perplexed*, part 3, chapter 51.
7. *Babylonian Talmud*, Section Sanhedrin 34a.
8. *The New York Times*, November 4, 1980, p. C 3.

CHAPTER 2

Stretching Time

■■■■■■■■■■■■■■■■■■■■■

Conflicts between science and religion result from mis-interpretations of the Bible.

—Maimonides

Absolute, true and mathematical time, of itself, and from its own nature, flows equably without relation to anything external.

—Sir Isaac Newton
Mathematical Principles of Natural Philosophy and His System of the World

Relativistic time $(T_2 - T_1) = (t_2 - t_1) [1 - \left(\dfrac{v^2}{c^2}\right)]^{0.5}$

—Albert Einstein
The Special Theory of Relativity

The heavens, the Earth, and all that they contain were created in six days. This is the fact that confronts us when we open the Bible. It is so totally at odds with every accepted scientific estimate of the age of the universe that most of us are tempted to read no further. Either we are almost forced to take the Bible as an allegorical statement of a past reality at best, or even more likely, as a mythological tale—an attempt to explain the unknowable. There are also those who believe that the events of the book of Genesis, taken at their most literal meaning, provide a true and accurate description of the universe's history. To them, the findings of archaeology and cosmology are seen as attempts to delve into questions best left to God.

Although science and theology offer divergent opinions on many aspects of our world, the difference in their statements of the age of the universe is particularly bothersome because it is a proven difference. The age of the universe has been measured using a variety of independent technological systems, including radio-active dating, Doppler shifts in starlight, and the isotropic "3° above zero" radiation background. The methods of these studies are totally unrelated. Therefore, an error that might have occurred in one would not appear in the others. Yet the data taken from these diverse studies present a strong and scientifically consistent argument for a very old Earth and an even older universe.

In sharp contrast to this range of tangible scientific evidence stands the faith that accepts the Bible's brief account of our genesis.

Any "proof" for or against the occurrence of the biblical Flood of Noah's time is weak. In Genesis we are told that the downpour lasted only 40 days and the resulting flood persisted for only 150 days. Sediments from so brief a period would probably not be extensive and, therefore, firm archaeological evidence may never be found.

What about Eve? Was she really taken from Adam's body some 5700 years ago as the Bible claims? In Genesis 2:21 it is recorded that God took Eve from Adam's "side" (*rib* is a mistranslation of the Hebrew; this is made abundantly clear by comparison with Exodus 26:20, where the identical Hebrew word is used for the *side* of the tabernacle). Whether rib or side or neither, the event is unlikely to be proven by physical evidence in archaeological finds.

But six days! That is an uncompromisingly explicit statement, and it refers to phenomena that are readily measurable by modern archaeological, paleontological, and cosmological instrumenta-tion. It is as if the first chapter of the Bible was deliberately included to make difficulties for people who want to believe bib-lical teachings. Rashi, commenting on Genesis 1:1, states that it was not necessary to start the Torah with the account of genesis. The Torah could have started with the events recorded in the book of Exodus.[1] The description of our genesis might have been omitted or at least relegated to some later book as an obscure, brief reflection on history. But no, there it is, right up front chal-lenging us in the very dimension that seems to rule our lives: the passage of time. No one is immune from the effects of its unre-lenting flow into the future. Some may question whether its flow can be slowed or stemmed, as did Einstein and, 200 years before him, Newton. But none of us can fail to feel the brevity of six days.

There is just no avoiding the issue. The Bible gives God six days to form mankind from the material produced at the creation. Current cosomology claims, it even proves, that nature took some 15 billion years to accomplish the same thing.

Which understanding is correct?

Both are. Literally. With no allegorical modifications of these two simultaneous, yet different, time periods.

It is unequivocal. Six 24-hour days elapsed between "the beginning," that speck of time at the start of the Big Bang, and the appearance of mankind, and, *simultaneously*, it took some 15 billion, 365-day years to get from "the beginning," or the Big Bang as astrophysicists call it, to mankind. We are not talking about easy explanations such as calling each day of Genesis 3 billion years so that 3 × 6 equals 18 billion years of cosmology. We also are not seeking changes in the functioning of the laws of physics. Such changes might include, for example, variations during the past millennia in the rates of radioactive decay of elements used in the dating of fossils. Vicissitudes of this nature would invalidate the methods of paleontology and archaeology, and so, the ages that those sciences attribute to prehistoric life would then be meaningless. We have never succeeded in altering rates of radioactive decay, and the temperatures or pressures required to alter nuclear processes (of which radioactive decay is one) are so extreme that, had they occurred, they would have obliterated the fossils themselves. There would be no fossil record! Radioactive carbon-14 had a 5600-year half-life in the early universe just as it has now. Theological arguments based on faulty understandings of either natural science or the Bible are counterproductive to a search for truth in any age and especially during an age that puts so much credence in science.

Misplaced fossils and changes in radioactivity are not needed to reconcile science and theology because the same single sequence of events that encompasses the time period from "the beginning" to the appearance of mankind did take six days *and* 15 billion years—*simultaneously*—starting at the same instant and finishing at the same instant.

Physics has proven this proposition to be rigorously correct.

In the following pages we will explore this seemingly illogical reality, bringing it to a logical conclusion.

THE BIBLICAL CALENDAR

Biblical dating is not totally at odds with archaeology. Within the total time span of the biblical calendar, we find no conflicts between its chronology and scientifically established dates for the entire post-Adam period, that is, the 57 centuries since Adam.

It is reasonable to suppose that if a calendar, which relates to a series of events, is found to be accurate for part of that series, then at least it is plausible that this same calendar may be accurate for the remainder of the events which it chronicles. When we compare the biblical dating of events that occurred after Adam with dates for these same events as ascribed by archaeological finds, there appear to be no conflicts; they match to within the accepted margin of error for the archaeological data. Only the early part of the Bible's calendar, which relates to events of the six days that precede Adam, appears to be in contradiction with the results of modern scientific inquiry.

There are many examples that can be used to establish confidence in the post-Adam biblical calendar. Here are several examples, early and late, biblically speaking.

Among the earliest post-Adam events of the Bible that might appear in archaeological finds is the invention of forged brass tools (Gen. 4:22). In the early Hebrew of the Bible, brass and bronze are denoted by the same word, nehoshet. The Bible attributes this development in forging to Tuval-Cain, a son of the biblical figure, Lemach. Archaeologically, we would call the time of Tuval-Cain the early Bronze Age.

To place a biblical date for the early Bronze Age, we must calculate the time that passed from Tuval-Cain's birth until the present. The genealogical line that leads to the present survived the Flood in Noah and his children. This line came from Adam through Seth (Gen. 5:3–32), Adam and Eve's third son, born to them after Cain murdered Abel (Gen. 4:8). The age of each generation of Seth's descendents is listed in the Bible and so it is possible to calculate the historical age during which they lived.

It is from these dates that we estimate the 57-century biblical age of the Earth.

Tuval-Cain was a descendent of Cain, not of Seth; perhaps because the line of Cain was destroyed in the Flood (Gen. 7), the Bible does not list the individual ages of Cain's descendents (Gen. 4:17–22). To learn the dates during which Cain's progeny lived, we compare their generations with the parallel generations of Seth. These dates are shown in Table 2. In this comparison, there are two fixed biblical reference points: the age of Cain at the birth of Seth, and the age of Noah at the Flood that destroyed Cain's progeny. In the year 1990, all the generations since Adam have a cumulative age of 5750 years. This biblical date for the dawn of recorded history is closely matched by the archaeological finds of the last two centuries. From 5750 we can subtract the time from Adam until Tuval-Cain (this is the biblical statement for the number of years that elapsed between the appearance of Adam and the invention of bronze casting) and learn the biblical date for the early Bronze Age.

In this calendar, we see that Adam was 130 years old when he and Eve had Seth. Seth was 105 years old when he begat Enosh, and so on until Noah. Tuval-Cain, we see from Table 2, was the last in the line of Cain to be born prior to the Flood. That puts

TABLE 2. PREFLOOD GENEALOGY

Genealogical Line of Seth	Age at Birth of Son (years)	Source in Genesis	Genealogical Line of Cain	Source in Genesis
Adam	130	5:3	Adam	4:1
Seth	105	5:6	Cain	4:1
Enosh	90	5:9	Hanoch	4:17
Qenan	70	5:12	Irad	4:18
Mahalalel	65	5:15	Mehuyael	4:18
Yared	162	5:18	Metushael	4:18
Hanoch	65	5:21	Lemach	4:18
Methusaleh	187	5:25	Yaval	4:20
Lemach	182	5:28	Yuval	4:21
Noah	600 at the Flood	7:6	Tuval-Cain	4:22
Total	1656		(brass tools)	

his life sometime in the lifetime of Noah. Using a midpoint of the age of Noah for the time that Tuval-Cain invented forging, we can place the early Bronze Age at approximately 1350 years after the appearance of Adam, or 4400 years (5750 − 1350) before the present.

The archaeology section of the Israel Museum houses examples of early brass tools. Their first appearance is dated at approximately 2400 B.C.E. or approximately 4400 years ago. It is of great interest for our discussions that during the last 100 years archaeologists have discovered that the early Bronze Age coincided with Tuval-Cain's lifetime, the biblical inventor of bronze tools.

For some people, the longevity of the pre-Flood generations is hard to reconcile with the 70- to 80-year span of modern life. Some modern scholars claim that the long lives refer not to one individual but rather to a combination of several lives into one for simplification of the biblical account. Nahmanides does not agree.[2] He claims that prior to the Flood, the conditions on Earth favored long life. The upheaval that accompanied the Flood changed the atmosphere and climate causing a gradual shortening of individual life spans. Regardless of which opinion is correct, we find that the dating of the Bible during the post-Adam, pre-Flood period matches the dating as found by archaeologists.

Similar examples of the accuracy of the biblical calendar are found throughout the ages encompassed by the Bible. In the book of Joshua (11:11–13) we are told that during the conquest of the land of Canaan, Joshua destroyed the city of Hazor, "burning it with fire. But as for the cities that stood on their hills, Israel did not burn any of them, except Hazor alone did Joshua burn." Later, in 1 Kings (9:15), we learn that Solomon rebuilt the towns of Hazor, Megiddo, and Gezer, using them as garrisons for his many chariots. Biblically, Solomon's work occurred in the tenth century B.C.E. It was 300 years after Joshua had destroyed the earlier Hazor on the same site. During the last century the ruins of Hazor were excavated. In the small hill formed by the gradual accumulation of soil and debris over an archaeological ruin, known as a tell (tel, in Hebrew), were found the charred ruins of the city that Joshua burned. Above this, but still buried within the tell, was the city built by Solomon. Hazor still shows the clear outline of the stables used for the horses of Solomon's chariots. The ruins are distinctly marked by the characteristic Solomonic shape for the gate of the city. The stables and the gate have the same shapes in each of the three cities mentioned in 1 Kings 9:15. Solomon may, in fact, have used the same architect for all three cities.

The biblical dating of Joshua's conquest of Canaan as the thirteenth century B.C.E. was scientifically confirmed by the definitive archaeological finds at Hazor during excavations between 1955 and 1968. Interestingly, this biblical date had been vigorously questioned by the famous British archaeologist John Garstang, in his "Joshua, Judges" (1931). Garstang, who was director of the British School of Archaeology in Jerusalem (1919–1926) considered the thirteenth century B.C.E. as being much too late to conform with his estimates of reality. Based on these archaeological finds, scientific opinion has changed and has come to agree with the traditional biblical view.[3] This is but one of so many examples of the truth literally springing from the Earth (Ps. 85:12).

The matching of the biblical and archaeological dates for the inception of the Bronze Age is doubly instructive specifically because it occurred in the post-Adam, but pre-Flood, period. It demonstrates the need to maintain a respect for both the natural and the Divine sciences. While helping to confirm the biblical calendar, it also dispels the reasoning that the Flood altered the fossil record. If the Flood had caused the claimed changes, then all pre-Flood events would be in error. Yet we have seen that the biblical date ascribed to the early Bronze Age, which occurred in that block of time *following Adam but preceding the Flood*, eminently matches the archaeologically established date for this same event. Both the archaeological record *and* the biblical record are valid.

So the biblical calendar for the time *after* Adam presents no problem to archaeology. But what about pre-Adam biblical chronology? Can we integrate the biblical data with that presented by cosmology and paleontology for the period extending from "the beginning" to Adam? Yes, we can, but only if we practice an exercise in logic that we refer to as "stretching time." This is the very heart of the matter.

How are we to stretch six days to encompass 15 billion years? Or the reverse, how do we squeeze 15 billion years into six days?

The suggestion is not as absurd as it may at first appear. In the Psalms of David we read, "A thousand years in Your eyes are as a day that passes" (Ps. 90:4). This presents the possibility that God's perception of time is quite different from mankind's. But the literal meaning of this verse is qualitative. It has the feel of time *seeming* to pass at different rates for different participants in an event, but not necessarily being different in reality.

This verse in Psalms is reminiscent of the dilation of time dealt with in Einstein's revolutionary thought experiments. Einstein demonstrated that when a single event is viewed from two frames of reference, a thousand or even a billion years in one can indeed pass for days in the other.

The basis we seek for matching the biblical and cosmological calendars is, in fact, partly found in relativity. But Einstein's theory is no longer a theory. It is an empirically observed reality: Einstein's *law* of relativity tells us that dimensions in space and the passage of time are not absolute. Their measurement is an intimate function of the relationship between the observer and the observed. It seems incomprehensible that the flow of time, which is so constant in our daily lives, can actually change—but it can.

SPECIAL AND GENERAL RELATIVITY

Among the most important aspects of modern physics that relate directly to our understanding of theology are the concepts of time: its origins and the absence of a unique, or constant and unchanging, measure for its passage. Because of the relevance of chronology to biblical interpretation, it is essential that we understand the insights that relativity provides toward our perception of the universe, its age, and its functioning.

Few theories have had such a profound effect on mankind's understanding of the world and of Genesis as the special and general theories of relativity. Prior to relativity, time was always considered to be absolute. Regardless of who measured the duration of an event, the elapsed time was thought to be the same. Some 300 years ago, Newton stated this belief eloquently: "Absolute, true and mathematical time, of itself, and from its own nature, flows equably without relation to anything external."[4]

Moreover, time and space were thought to be unconnected—one did not influence the other. After all, what other connection could there be between the distance separating two points and the flow of time other than the fact that a greater distance requires a longer time to traverse; pure and simple logic.

The theories put forth in Einstein's special theory of relativity (1905) and then in his general theory of relativity (1916) changed this understanding of space and time as profoundly as the light of a lamp changes our perception of a once-darkened room.[5]

The long trek toward Einstein's insights started in 1628, when Johannes Kepler[6] observed a curious phenomenon. He noticed that the tails of comets always curved away from the Sun. A shooting star streaking through the heavens invariably has its tail flaring out behind it. This is also the case for a comet during its approach to the Sun. But after the comet passes the Sun and starts its return flight to the outer reaches of the solar system, the situation changes dramatically. The comet's tail takes the lead and streams ahead of the main body. This position dramatically contradicts the usual concept of a tail! Kepler postulated that the location of the comet's tail relative to its main body was the result of pressure exerted on it by the sunshine. The tail, being less dense than the comet, is more subject to the forces of the solar radiation than is the main comet body. The solar radiation, in effect, blows the tail, always pushing it away from the Sun. If it were not for the gravitational attraction of the main body of the comet, the fine particles of the tail would be blown entirely away. Kepler's observation was the first indication that radiant energy, such as light, might also have a mechanical (pushing) force associated with it. This was an important concept because it meant that light, which had always been assumed to be substanceless, might also have a mass or weight. It was 273 years later, in 1901, that the pressure of radiant energy was first measured. E. F. Nichols and G. F. Hull, shining a powerful light on a mirror suspended in a vacuum, measured a deflection of the mirror caused by the pressure of the light. It was the laboratory analogy of the sunlight pushing the comet's tail.

In 1864, working with discoveries made by Michael Faraday in electricity and magnetism, James Clerk Maxwell postulated that light and all other electromagnetic radiation moved as waves at the same fixed speed through space.[7] The microwaves in the kitchen microwave oven, the light by which we read, the X rays that let a doctor see a broken bone, and the gamma rays released in a nuclear explosion are all composed of electromagnetic radiation. The only difference among them is their wavelength and the frequency of the wave. The higher the energy of the radiation, the shorter its wavelength and the higher its frequency. Aside from this they are identical.

In 1900, Max Planck proposed a description of that electromagnetic radiation that was a radical departure from all previous theories. It had previously been assumed that the energy radiated by a warm object, such as the warming glow of a red-hot iron, was released in a smooth and continuous process. This was thought to continue until the heat had dissipated and the object returned to its initial condition, and this certainly appeared to be true for the cooling of a hot iron to room temperature. But Planck observed that just the opposite was true. Instead of being a smooth, continuous flow, the energy was released in discrete units, as if a glowing iron gave off its heat by shooting out a rapid stream of tiny hot pellets.

Planck theorized that these pellets were actually the radiant energy. He called these pellets "quanta," and thus quantum mechanics was born. Because all radiation moved at the same speed (the speed of light), the quanta must do likewise. Although the quanta all moved at the same speed, they might not all have equal energies. The energy of an individual quantum, he proposed, was proportional to its frequency of vibration as it sped through space, similar to a tiny rubber ball that is continuously flexing in and out as it speeds along its path. In the visible range, our eyes can measure the frequency at which the quanta pulse. We call that measure the color of the light. It is because of this quantized release of energy that an object, when gently heated, first glows red, then, as its temperature rises, emits additional colors, going through the rainbow spectrum of increasing energy and frequency, resulting in a mix of all frequencies. This we see as white heat.

Here lies a paradox—the very theory that described light as being composed of particles, called quanta, also described the same energy in terms of frequencies (see Figure 1). But frequencies are associated with waves and not particles. Furthermore, the speed of the light was always to be constant. But what would happen if the object emitting the light or the subject observing the light were moving? Would the speed of their motion add to or subtract from the speed of the light? Logically it must, but then the speed of light would not appear to be constant! The pressure that light exerts on the tail of a comet or on the mirror in the Nichols-Hull experiment implies a change in momentum of the light as it crashes to a stop on striking a surface. That is the source of pressure from any moving object. The spray of water from a hose pushes a ball along the ground because the water has mass and this mass has a velocity that is stopped when the water hits the ball. In stopping, the water transfers its momentum to the ball and the ball rolls away. The very definition of momentum—equaling the mass (m) or weight of an object times its velocity (v), or mv—requires that the moving light have a mass. Somehow these wavelike particles of light had a mass associated with them, even though there was no material residue of this mass left on the surface on which a light was shining. There was no mess to wipe up after a light spilled onto a surface. To this day, we are still searching for a unified theory that will fully explain these phenomena of light and all radiant energy.

Concurrent with the investigations into the nature of radiant energy were studies relating to the propagation of light. It was logical that because light and other electromagnetic radiation were in a sense waves, they required a medium for their wavelike transmission. They could not propagate through a vacuum. Just as sound requires some form of a material substance such as air to carry its wavelike energy, so it was assumed that light also needed its own special substance for transmission. To account for the transmission of radiant energy through space, such as the light and warmth of the Sun reaching the Earth, it was postulated that the universe must be permeated by an invisible, insensible medium, called ether. Ether was assumed to be present even in the vacuum of space.

Because the light would travel within this ether, the paradox of its constant speed was solved. Light would travel at a constant speed, not in relation to the speed of the emitter of the light and not even in relation to the speed of the observer of the light. Light

Figure 1. *Light as a wave and as a particle.*

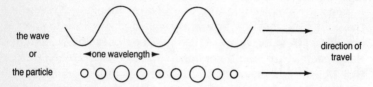

the wave

or

the particle

◄one wavelength►

direction of travel

The dual nature of light is such that when we measure light by passing it through a narrow opening, the light behaves just like an ocean wave as the wave passes through the narrow opening of a harbor. Both the light wave and the ocean wave, after passing through the opening, spread out on the other side of the opening in the form of a crescent. On the other hand, when measuring light by having it shine on a certain metal, it behaves as a stream of tiny particles hitting the metal. The light knocks out electrons from the metal, one at a time, just as tiny pellets shot into a paper target would punch out bits of paper as each pellet strikes. The energy of the light wave is determined by the wavelength. The energy of the light pellet is determined not by its speed, but rather by the frequency at which the pellets of light, the photons, pulse as they travel along at the speed of light.

would travel at a constant speed in relation to this universally present ether. Observers moving through the ether might see the light moving faster or slower depending on their direction of motion relative to the light, but the speed of light relative to the fixed ether would remain constant. This would be the same as the propagation of sound. In still air, sound moves through the air at a constant velocity, about 300 meters per second (about 600 mph) at sea level, regardless of whether or not the object making the sound is moving. The sonic boom of a plane traveling faster than 300 meters per second is in effect the noise of the plane crashing through and passing its own sound waves. Here the producer of the sound, the plane, is actually going faster than the sound it is making.

In the debates within the scientific community over the nature of the yet-to-be-observed ether, no one yet suspected that the flow of time could be related to the flow of light, but this enlightenment was soon to be brought to man's attention.

In 1887, Albert Michelson and Edward Morley published the results of an attempt to measure the effects of this theoretical ether.[8] They compared the elapsed time for light to make a round-trip of equal distances, perpendicular and parallel to the motion of the Earth as it orbits the Sun. Because the Earth orbits the Sun at approximately 30 kilometers per second, we must be pushing through the ether at this same speed. If radiant energy followed the laws of all other waves, this must cause a change in the travel time of the light used in their experiments. The effect of the Earth's motion through the ether should be no different than the effect of a stiff wind carrying sound.

To the astonishment of all concerned, Michelson and Morley did not record any effect of this 30-kilometer-per-second speed. The original experiment and later, technically improved versions of the same experiment revealed a most extraordinary situation: There is no effect of the Earth's motion on the speed of light. ✓

This was very confusing. We have a speed of light (c) that is constant at 299,792.5 kilometers per second whether or not the source of the light or the observer of the light is moving. To add to the confusion, the same light beam behaves like a wave when measured one way and behaves like a particle when measured another way. It is as if we were standing on a jetty watching waves rolling in from the ocean, and at the blink of an eye, the usual crests and troughs change into a stream of individual balls of water, pulsing as they flow through the air just above sea level. With another blink, the waves return, and the balls are gone.

In 1905, in the midst of the confusion, Albert Einstein entered, uniting the principle of relativity with the theory of relativity. During that year, Einstein published a series of papers that literally changed mankind's understanding of our universe. Five years earlier, Max Planck had proposed the quantum theory for light. Using Planck's theory, Einstein was able to explain an interesting phenomenon. Light shining on certain metals releases electrons and in so doing creates a potential for the flow of electric current.[9] Einstein postulated that this "photoelectric" effect was the result of quanta of light (photons) literally knocking electrons out of their atomic orbits. The photon, it turns out, has mass while in motion (recall that it moves with the velocity of light, c), but has no mass when stopped (zero rest mass). It has properties of both a particle—it occupies space and while in motion has a given mass and so, as Kepler suggested, it can push things around, like the tail of a comet—and a wave—it has a frequency of vibration that is proportional to its energy. Matter and energy seemed to be intimately connected in the photon. Einstein discovered this relation and described it in his now-famous formula: $E = mc^2$. The energy of a photon, E, equals its mass, m (while in motion), times the speed of light squared, c^2. Einstein saw that this relation was applicable to all masses and all forms of energy. It is the basis of the special theory of relativity.

Integrating this concept into one's thought processes is quite an exercise in mental gymnastics. For example, take a given object. The mass, or weight, of the object while at rest is called, in technical terms, its rest mass. Now give the object a strong sustained push. It acquires velocity and in so doing acquires kinetic energy proportional to the velocity. But because the E of $E = mc^2$ refers to all forms of energy, the object's total energy is represented by its rest mass plus its kinetic energy (the energy of its motion). Einstein's relation demands that the mass of the object actually increase as the object's velocity increases.

The essence of relativity tells us that when the velocity of an object changes, its mass also changes. At small velocities, the mass of an object is essentially the same as the original rest mass. That is why for day-to-day activities Newton's descriptions of natural laws are quite exact. For galaxies streaming through space or subnuclear particles speeding within an accelerator, the story is entirely different. Both may have relative velocities that are a large fraction of the speed of light. And furthermore, the increases in mass are observable only by someone at rest relative to the moving object. An observer moving along with the object cannot detect this change in mass.

As we shall learn later, this interchangeability between energy and mass is discussed eloquently by Steven Weinberg in his book *The First Three Minutes* and by Nahmanides in his *Commentary on Genesis*. Both refer to the mass-energy duality in their descriptions of the very early universe.

The special theory of relativity is based on two facts: the principle of relativity and the constant speed of light. The principle of relativity, postulated by Galileo Galilei 300 years ago, was updated by Einstein. It states that all the laws of physics (which are really no more than the laws of nature) are the same in all unaccelerated systems: those systems having smooth, uniform motion. Such systems, in the jargon of physics, are called *inertial reference frames*.

A reference frame is how you see yourself in relation to the world. At times, it is not inertial, such as when we are rocking in a rocking chair looking at a stationary object. The principle of relativity tells us that as long as we are within an inertial reference frame we cannot use the laws of physics to determine the frame's motion because that motion has no effect on any of the dimensions that we measure within the frame. This is why there is no sensation of motion while flying at a constant speed in calm weather. On the other hand, the constantly changing motion of the rocking chair makes our system noninertial and so we can feel our movement.

We all have experienced the inability to measure absolute motion. At a stoplight, the car next to us slowly rolls backwards. Or are we rolling forward? At first it is difficult to tell. Our train gently rolls out of the station. We awake from dozing and notice that the train next to us is moving backwards. Or at least we believe this to be the case. Until our reference frame, our car or train, starts to accelerate (and ceases its inertial motion) it is not clear what is in motion and what is at rest.

It may seem contradictory that Einstein told us that the mass of an object is a function of its velocity and now we say that we can't determine our motion by measuring its change in mass. But there is a subtle difference. Measured within the inertial frame, all values remain the same. Measured across reference frames (i.e., from a frame that is in motion relative to another) the values of size and mass change. If all the parts of the universe were in equal and uniform motion, relativity would have little relevance for our present study. But this is not the case. It is the experience of observing events across reference frames that will be essential when we return to the biblical analysis of cosmology.

The second foundation of the special theory of relativity is less accessible. In fact, it is outrageous. It states that the velocity of light, c, is constant ($c = 2.997925 \times 10^8$ meters per second in a vacuum—always) within all reference frames and *across* all reference frames. This fact was hidden within the results of the Michelson-Morley experiment. If you consider the implications of this statement, you will realize the audacity of such a concept. What Einstein had the courage to state was that regardless of the velocity of the observer, toward or away from a source of light, the speed of that light remains c. No other form of motion (e.g., sound waves) has this property. It defies all logic.

If the pitcher throws a baseball to the catcher at 90 mph, the catcher sees the ball approach him at 90 mph. Now if, contrary to the rules, the catcher runs toward the pitcher at 20 mph, the speed of the ball relative to the catcher is 110 mph (90 + 20). The speed relative to the pitcher remains at 90 mph. For the next pitch, instead of throwing a ball, the pitcher flashes a picture of a ball toward the catcher. The light of the picture moves out from the pitcher's hand at the speed of light (1.0 c), or approximately 300 million meters per second. The quick catcher runs toward the pitcher at a tenth the speed of light, or 30 million meters per second. So what does the catcher see? Light approaching him at 330 million meters per second (1.1 $\times$ c)? No! That is the whole paradox of light—mind-twisting, annoying, at times infuriating, but for us emancipating. The catcher sees the light approaching him at exactly the speed of light, 300 million meters per second, even though he is running toward the light and so in effect adding his speed to the speed of the light. Light, regardless of the velocity of the observer relative to the light source, is always observed to move at c. Always. And at what speed does the pitcher, who is standing still, see the light move out from him? Right, also 1.0 c. How can two observers, one moving and one standing still, measure the same speed for the light? Common sense and logic say that they cannot. But relativity says they do. And the Michelson-Morley experiment proved it.

Both see the same speed of light because fundamental to relativistic mechanics, and to the laws of the universe in which you and I dwell, is the fact, incomprehensible as it may be, of mass, space, and time dilation. This dilation is just enough so that within a given reference frame nothing absurd seems to be happening. There is no perception of the dilation; only when looking in from the outside is it observed. Looking into a reference frame moving by us, we would observe that dimensions of an object along the direction of motion are contracted relative to the same object's dimensions when at rest. Moreover, the clock that had worked precisely while at rest is now running slow relative to a clock at "rest" in our reference frame.

Combining the principle of relativity and the constancy of the speed of light leads inescapably to the phenomenon of time dilation. We can demonstrate time dilation by a thought experiment, the same kind of *Gedanke* ("thought") experiment used by Einstein to develop the conceptual framework for his work in relativity. Taylor and Wheeler, in their classic book *Spacetime Physics*,[10] present a version of the following example of these mental experiments.

Let there be two reference frames, one at rest and one in motion. The one at rest is a standard physics laboratory. The other is a high-speed, transparent, totally permeable rocket ship with permeable and transparent scientists aboard. The rocket can pass right through our laboratory without disturbing it. In the lab, we flash a light that leaves point A and travels diagonally to a mirror (M in Figure 2). From the mirror the light is reflected diagonally back to point B. We time the arrival of our rocket ship so that point A in the lab coincides with point A in the rocket at the very instant the light flashes. We adjust the speed of the rocket so that

Figure 2. *A flash of light observed simultaneously in a stationary laboratory and in a speeding rocket.*

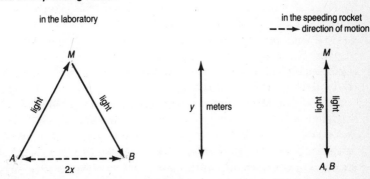

point A of the rocket passes point B of the lab at the exact instant that the light flash reaches lab point B. From inside the rocket, it appears as if the light flashed from rocket point A, traveled straight up to rocket point M, and returned directly to rocket point A. Because the rocket's speed is constant (inertial), the people inside have no idea that they are moving.

The distance traveled by the light as viewed in the rocket was straight up and down, or 2y meters. The distance traveled by the light as viewed in the laboratory was the triangular path A to M to B. The triangular path in the lab must be longer than the direct path taken in the rocket. We see this qualitatively by comparing the light paths shown in the diagram. We could calculate the difference in the travel distance rigorously by the Pythagorean theorem. In the rocket, the path of travel for the light is shorter than the triangular route taken by the light as observed in the lab.

Recall that in both systems the speed of light is the same. This is one of the proven fundamentals of relativity. Now, in all cases, the time taken for a journey is distance traveled divided by velocity of travel. The time to travel 100 miles, when moving at 50 mph, is 2 hours. Because the speed of light for both the scientists in the lab and those in the rocket is the same, c, and the distance traveled by the light is longer in the laboratory than in the rocket, the elapsed time between the light flash leaving A and its reception at B *must have been longer* in the lab than in the rocket.

It was a single event. There was only one flash of light and one passage of light that was observed in the two frames. Yet the duration of this same event was actually different when measured in the two separate time frames.

The difference in perceived time is called relativistic time dilation, the dilation that makes the first six days of Genesis reassuringly compatible with the 15 billion years of cosmology.

The concepts related to general relativity flow from those of special relativity but are more complex. While special relativity deals only with inertial systems, general relativity takes into account both inertial and noninertial (accelerated) systems. These include situations in which forces such as gravity are influencing the motion of objects. The aspect of gravity that is relevant to our study is the extraordinary fact that, just as with speed, gravity also causes time dilation. A clock on the Moon runs more rapidly than the same clock when on the Earth because the Moon has ✓ less gravity than the Earth. As we shall see, gravity is a key factor in the reconciliation between Genesis and the Big Bang.

Forces of gravity are felt in the identical way that forces that cause acceleration are felt. For example, the force felt in a rising elevator is the floor pressing against your feet as, in effect, it pushes you up along with the elevator. This feels the same—and, in fact, is the same—as the force you would feel in a stationary elevator that suddenly, somehow, became subject to an increase in the gravity from beneath. As Einstein reasoned, because gravity feels the same as any other force that induces motion, it must have the same effect as those forces. Because accelerating forces produce motion and time dilation, so must changes in gravity induce time dilation.

TIME DILATION

Because the aspect of relativity related to time dilation is essential to the unification we seek between the cosmological and biblical calendars, it is essential to demonstrate that time dilation actually occurs. Relativistic changes become significant only when relative velocities approach the speed of light. Even at a tenth the speed of light, which is 30 million meters per second, the dilation is less than 1 percent.

Velocities close to the speed of light are rarely encountered in daily life but are standard fare in cosmology and high-energy physics laboratories. The fact that we can actually measure time dilation does not make the phenomenon any less difficult to grasp conceptually. It does, however, remove it from the pigeonhole of pure theory and place it in the realm of empirical fact. Demonstrations of time dilation have come from environments ranging from high-energy physics labs to commercial flights of scheduled airlines.

One of the many elementary particles produced in physics laboratories is the mu-meson. It decays with a 1.5 microsecond half-period. Mu-mesons, in addition to being products of experiments in high-energy physics laboratories, are also produced near the top of the Earth's atmosphere as cosmic rays slam into nuclei of atmospheric gases. There, the energy of the incident cosmic radiation is so great that, when formed, the mu-mesons have a speed that is almost the speed of light. At such high speeds, they experience measurable time dilation. Even moving at close to the speed of light, 200 microseconds are required to travel the 60 kilometers from the altitude where the mu-mesons are formed to the Earth's surface. Because the mu-meson has a 1.5 microsecond half-period, this 200 microsecond travel time uses 133 of its half-periods. Recall that in each half-period, half of the remaining particles decay. After 133 half-periods, the fraction of mu-mesons that should remain and reach the Earth's surface is $\frac{1}{2} \times \frac{1}{2} \times \frac{1}{2} \ldots$ repeated 133 times or less than 1 millionth millionth billionth billionth billionth of the mu-mesons that started their journey downward toward the Earth. This is a number so small that almost no mu-mesons should ever reach the Earth's surface. Most should have decayed in transit. Yet when the number of mu-mesons being produced at the top of the atmosphere is compared with the number reaching the Earth's surface, we find to our surprise that $\frac{1}{8}$ have reached the Earth's surface.[11] The survival of $\frac{1}{8}$ of the mu-mesons means that during their 60- kilometer trip only three half-periods elapsed: $\frac{1}{2} \times \frac{1}{2} \times \frac{1}{2} = \frac{1}{8}$. To the mu-meson traveling at close to the speed of light, the elapsed (relativistic) time is only three half-periods: 4.5 microseconds (3 × 1.5 microseconds). To the observer on the ground at least 200 microseconds passed because that is the shortest possible time needed to travel the 60 kilometers from the top of the atmosphere to the Earth. For the *same, single event* two very different times elapsed: 4.5 microseconds in the time frame of the speeding mu-meson and 200 microseconds in the time frame of the observer on the ground. Remember, it is one event. Due to the relative motion between the observer and the observed, two very different times elapsed. Both are absolutely correct!

But mu-mesons are exotic particles and a skeptic could shake his head in humored disbelief. After all, no observer actually traveled with the mu-mesons. We relied on their half-period to be the clock that traveled with them.

How about a real clock and human observers traveling with the clock to measure the effects of time dilation directly? That should be convincing. And that is just what is reported in the prestigious journal *Science* by scientists Hafele and Keating of Washington University and the U.S. Naval Observatory.[12] They sent four cesium-beam clocks on around-the-world trips aboard commercial, scheduled TWA and Pan Am flights using Boeing 707 and Concorde aircraft. Cesium-beam clocks were used because of their extreme precision.

The Earth rotates from west to east. Viewing the Earth from space high above the north pole, we see that on the eastward flight, the speed of the plane added to the speed of the Earth. As predicted by relativity, the flying clocks lost time relative to cesium-beam clocks stationed at the U.S. Naval Observatory in Washington, D.C. (the source of all the clocks in this study). On the westward flight, the speed of the plane subtracted from the Earth's rotation and the clocks, as predicted, gained time. In the words of Hafele and Keating, "In science, relevant experimental facts supersede theoretical arguments. These results provide an unambiguous empirical resolution to the famous clock paradox."[13]

Not just the sensation of time, but the actuality of time's passage changes in accord with the relative motion of the observers. Within each of the given reference frames all seems natural. But when the frames come together and the clocks are compared, the passage of time, the actual aging, is seen to have been different.

Especially exciting about the Hafele-Keating time dilation experiment is that it tested both special and general relativity. General relativity demands that differences in the force of gravity affect dimensions just as a difference in relative speed in special relativity does. The force of a gravitational field on any object is inversely proportional to the square of the distance between the objects. If we double the distance of separation, the gravitational attraction drops fourfold. The farther one is from the Earth, the less the pull of the Earth's gravity. Because the planes were high above the surface of the Earth (the 707s flew at 10 kilometers altitude and the Concord at 20 kilometers), their experience of the Earth's gravitation was different than that of the clocks remaining on the Earth's surface at the Naval Observatory. The changes in the clocks' time matched the predictions of general relativity (which includes the effects of motion plus gravity).

This experiment and all other similar experiments have proven

Einstein's special and general theories of relativity to be true descriptions of reality in our universe. The theory of relativity is no longer a theory. It is a proven, empirically observed fact. Thus the theory of relativity has become the law of relativity.

With this background in relativity, one of the natural sciences of our universe, let us continue our discussion of the first six days of Genesis, a time when natural and Divine sciences at first glance seem to disagree.

STRETCHING TIME

Let us now consider changes that have occurred since that instant of "the beginning," changes in relationships among the Creator, the universe, and man. In doing this, we must keep in mind that we can experience differences in time's passage only when comparing events *across* the boundary of two reference frames. And even then, only when the gravity of two reference frames is vastly different, or when the relative velocities between the two reference frames approaches 300 million meters per second, the speed of light. *Within* a reference frame, regardless of its relative speed or gravitational force, all will seem perfectly Newtonian, which means, perfectly normal and logical, just as it does for all of us even though at this moment we are hurtling through space.

The Creator had and has a stake in the universe. We can assume that from the fact that the universe exists. What the stake is, however, we don't know. We can get some hints, though, from interactions that have occurred between the Creator and the universe. Traditional theology maintains that had the Creator wished to form the universe in a single act, then that Creator could have done so. From the biblical narrative, it is clear that the plan was not to bring a ready-made universe into existence in a single stroke. For some reason a gradual unfolding was chosen as the method. Instead, the first two chapters of Genesis describe the universe's step-by-step formation.

If you play by the rules under which the universe now functions —the physical laws we know today—then the gradual development of the universe from the stuff of the Big Bang was necessary to bring about the development of man. But the Earth and all that dwell on it are not *direct* products of the Big Bang. We are told explicitly that in the beginning the Earth was void and unformed or, in the Hebrew, *tohu* and *bohu*. Leading particle physicists now refer to *T* and *B* (*tohu* and *bohu*) as the two basic building blocks of *all* matter. The forces of the Big Bang literally pressed this *T* and *B* into hydrogen and helium—almost no other elements were formed at that time. Instead, it took the alchemy of the cosmos to convert this primordial hydrogen and helium into the rest of the elements.

The Earth and the solar system are a mix of matter that has come down to us after uncounted cycles of supercompression within the cores of stars. This compression squeezed the hydrogen and helium so tightly that their nuclei combined and recombined to form the heavier elements such as carbon (literally the atomic staff of life), iron, uranium, and the other 89 elements that comprise the universe. These stars then exploded and spewed their newly formed elements into a universe hungry to reuse them in the forming stars. The births of stars and their deaths were all needed to recook the hydrogen and helium formed in the first few moments following the Big Bang into the elements needed for life as we know it. Biblical commentators such as Maimonides and Rashi expounded traditions that God created and destroyed many worlds during the process of establishing life on Earth. But my source of information here is not Maimonides; it is the astrophysicists Woosley and Phillips.[14]

Now if there were only six days prior to Adam, how do we squeeze all the cycles of formation and destruction of worlds into the allotted time? The biblical commentators on whom we rely said explicitly that the first six days of Genesis were six 24-hour days. This means that whoever was in charge recorded the passage of 24 hours per day. But who was there to measure the passage of time? Until Adam appeared on day six, God alone was watching the clock. *And that is the key.*

During the development of our universe and prior to the appearance of mankind, God had not yet established a close association with the Earth. For the first one or two days of the six days of Genesis, the Earth didn't even exist! Although Genesis 1:1 says "In the beginning God created the heavens and the earth," the very next verse says that the Earth was void and unformed. The first verse of Genesis is a general statement meaning that, in the beginning, a primeval substance was created, and from this substance the heavens and the Earth would be made during the subsequent six days. This is explicitly stated later in Exodus 31:17: "For six days God *made* the heavens and the Earth." From what were the heavens and the Earth "made" during these six days? From the substance created "in the beginning" of those six days. Because there was no Earth in the early universe, and no possibility of an intimate tie or a blending of the reference frames, there was no common calendar between God and the Earth.

Relativity has taught us that there was not even a Divine option of choosing a calendar that would eventually be compatible with *all* the various parts of the universe, or even with the more limited number of parts that eventually contributed to mankind. The law of relativity, one of the inherent traits of the universe established at its creation, makes it *impossible* for a common reference frame to have existed between the Creator and each part of the mix of matter that eventually became mankind and the Earth on which we stand.

According to Einstein's law of relativity, we now know it is *impossible* in an expanding universe to describe the elapsed time experienced during a sequence of events occurring in one part of the universe in a way that will be equal to the elapsed time for those same events when viewed from another part of the universe. The differences in motions and gravitational forces among the various galaxies, or even among the stars of a single galaxy, make the absolute passage of time a very local affair. Time differs from place to place.

The Bible is a guidebook for mankind's passage through life and time. To instill in mankind a reverence for the physical wonders of the universe, this guide includes a description of the development that led from a void, unformed universe to a home suitable for mankind. But choosing an all-encompassing time frame to describe that span of development is nearly impossible because so many factors have an intimate and profound effect on the rate at which time passes. These include the forces of gravity within the multitude of stars that converted the primeval hydrogen and helium into the elements of life, the motions of the intergalactic gases as they contracted into nebulas and then into stars, the supernovas' explosions marking the death and ultimate rebirth of the stars from which the Milky Way formed, and the mass of the Earth. The flow of time had been the one aspect of life that, until Einstein's insight, we were so erroneously certain was constant. It is unrealistic, no it is impossible, for a single clock to have timed all the ages of all the cosmic stuff of which we are composed.

The odyssey that stretched between the stuff of the Big Bang and the matter of today was too complex, too varied, to be timed by a single clock. Who can say how many galaxies or which supernovas contributed to the elements that comprise our physical bodies? We humans, and everything else in the solar system, Sun and planets included, are the debris of bygone stars. We are literally made of star dust. To which atoms of carbon, nitrogen, or oxygen would the time relate? Yours or your neighbors'? Those which constitute a speck of your skin or those found in a drop of your blood? It is possible that each may have originated in a separate stellar core and therefore has its own unique age. Until the formation of the Earth, the processing of the cosmic stuff of which we are composed occurred in a myriad of stars, concurrently and sequentially. Each star and each supernova had its own gravity, its own speed, and so its own space-time reference frame.

A billion cosmic clocks were (and still are) ticking, each at its own, locally correct rate. Universally, they all started at the Big Bang and, at the very same instant, reached the moment when Adam appeared. But the absolute, local time that passed from "the beginning" to the instant of their particular contribution to humanity was very different for each star and so for each contribution of matter. Although the processing started and concluded at the same times, the legacy of Einstein proves for us that the age of each bit of matter was very different from the other bits of matter with which it ultimately mixed to form the solar system and then mankind. Our reasoning is no more and no less subtle than finding the 200 microseconds in the 4.5 microseconds that elapsed while the mu-mesons, created by cosmic rays striking the top of our atmosphere, traveled to the Earth's surface. In 4.5 microseconds, 200 microseconds elapsed. We can better understand this proven fact with the help of Einstein's thought experiment, in which the scientists aboard a speeding rocket and those in a stationary laboratory measured two very different times for a single event. This has no similarity to the claim by the late W. C. Fields that one night he spent a week in Philadelphia. His was an emotional sensation; ours is a physical fact. When we talk of a billion years, we don't mean it *felt* like a billion years. It *was* a billion years! If, during those first six days, a clock had been suspended in that part of the universe now occupied by the Earth, it would not necessarily have recorded 15 billion years. In the early universe, the curvature of space and time in this spot was probably very different from what it is now.

Instead, a compromise had to be made to describe the sequential development of the universe. This compromise was to choose, for the time preceding Adam, the Creator's own reference frame that viewed all the universe as a single entity.

The formation of Adam was qualitatively different from all other events following the creation of the universe. It signaled a monumental change in God's relationship with the universe. We know that all entities of the universe, organic or inorganic, living or inert, are composed of matter, the origin of which can be traced back to the original creation. In this respect, mankind is no different. We are told explicitly that our material origins lie in the "dust of the ground." All animals (Gen. 1:30) including man (Gen. 2:7) were given a soul of life (in Hebrew the *nefesh*). However, Adam alone was given something new, unique in the universe—the living breath of God (Gen. 2:7).

It is only at the instant when God places in Adam this breath (in Hebrew the *neshamah*), that both the created and Creator become inseparably linked. It is at this juncture that one out of billions of possible clocks was irrevocably chosen, by which all future acts would be measured.

In the jargon of relativistic physics, it was at the moment of Adam's appearance that the part of the universe where man dwells started to operate in the same space-time reference frame as its Creator. At this point, the chronology of the Bible and the flow of time on Earth became one—the common space-time relation between God and man was now fixed.

The result of this new linkage is at once apparent from the biblical text. There is a parity between ages that the Bible ascribes to *post*-Adam events and corresponding archaeological estimates of dates for the same events. The Bronze Age of the biblical calendar and the Bronze Age of archaeology do coincide. Hazor was destroyed by Joshua 3300 years ago according to the Bible and now, after much research, according to archaeology. The post-Adam part of the biblical calendar makes sense to us and the discoveries of the Dead Sea scrolls prove that the Bible predates by thousands of years any of the modern archaeological finds that confirm it. Had we not learned the law of relativity and had we timed the passage of events on Earth during this post-Adam period from a different vantage point in the universe, we would now be wondering why our perceptions of these times are so different from those recorded by the clocks of the Earth.

In the first six days of our universe's existence, the Eternal clock saw 144 hours pass. We now know that this quantity of time need not bear similarity to the time lapse measured at another part of the universe. As dwellers within the universe, we estimate the passage of time with clocks found in our particular, local reference frame; clocks such as radioactive dating, geologic placement, and measurements of rates and distances in an expanding universe. It is with these clocks that humanity travels.

When the Bible describes the day-by-day development of our universe in the six days following the creation, it is truly referring to six 24-hour days. But the reference frame by which those days were measured was one which contained the total universe. This first week of Genesis is not some tale to satisfy the curiosity of children, to be discarded in the wisdom of adulthood. Quite the contrary, it contains hints of events that mankind is only now beginning to comprehend.

Biblical sages long ago warned us that our perception of the events of the first six days of Genesis would be inconsistent with our understanding of nature for the time following Adam. They learned this from the descriptions of Sabbath rest contained in the Ten Commandments. If we were to compare the text in Exodus 20:11 with Zechariah 5:11 and 2 Samuel 21:10, we would find that the same word for resting is used. The usage in these texts reveals that the intent is not that God "rested" on the Sabbath. Rather the Creator caused a repose to encompass the universe that had been made during the first six days. Our perception of this repose, according to Maimonides,[15] is that from this first

Sabbath and for all thereafter, the laws of nature, including the flow of time, would function in a "normal" manner. In contrast, the flow of events that occurred *during* the first six days would appear illogical, as if the laws of time and nature had been violated. The sages' predictions of a perceived incongruity between the biblical and scientific views of the early universe have, in fact, been met.

The first Sabbath marks the start of the post-Adam calendar. It is this portion of the biblical calendar that satisfies our perceptions of reality based on logic. The extraordinary *fact* of the relativity of time, Einstein's law of general relativity, has extended the validity of the biblical calendar into those first six days. It has obviated the need to explain fossils as being placed in our world by the Creator to test our belief in Genesis or to satisfy our curiosity. Radioactive decay in rocks and meteorites and fossils accurately records the passage of time, but the passage as it was and is measured by the clocks of our earthbound reference frame. That time was, and still is, only relatively, that is only locally, correct. Other clocks in other reference frames record earthbound events at very different, but equally correct, times. This will always be the case as long as the universe follows the laws of nature.

We can only guess at what the perception of the flow of time in the early universe would have been for a human privileged to view those events from the eternal reference frame of our Creator. I imagine Genesis 1:31 gives us a hint of what that perception would have been: ". . . and there was evening and there was morning, the sixth day."

Notes

1. Rashi, *Commentary on the book of Genesis*, 1:1.
2. Nahmanides, *Commentary on Torah*, Genesis 5:4.
3. Thomas, ed., *Archaeology and Old Testament Study*.
4. Newton, *Mathematical Principles of Natural Philosophy*.
5. Einstein, *Relativity: The Special and General Theories*.
6. Cohen, *The Birth of a New Physics*.
7. Pagels, *Perfect Symmetry*.
8. Shankland, "The Michelson-Morley experiment," *American Journal of Physics*, 32 (1964): 16.
9. Hermann, *The Genesis of the Quantum Theory (1899–1913)*.
10. Taylor and Wheeler, *Spacetime Physics*.

11. Ibid.
12. Hafele and Keating, "Around-the-world atomic clocks: observed relativistic time gains," *Science* 117 (1972): 168.
13. Ibid.
14. Woosley and Phillips, "Supernova 1987A!" *Science* 240 (1988): 750.
15. Maimonides, *The Guide for the Perplexed*, part 1, chapter 67.

CHAPTER 3

Inklings of Expansion
A Big Bang in a Grain of Mustard

■■■■■■■■■■■■■■■■■■■■■■■

*Don't make yourself overly wise; why should you destroy
yourself?*

—Ecclesiastes 7:16

We have found within the six days of Genesis the time to
encompass the billions of years of cosmology. The calendar as we experience it begins only with the appearance of man. For dates prior to this, we found it necessary
to leave our Newtonian view of the world and to enter Albert
Einstein's universe of relativity. With this relativistic understanding of time, let's return to the very beginning of time and go step
by step through those first six days of creation (or 15 billion years,
if that is your preferred reference frame). In the light of Genesis
(1:1), we can better understand the history of our universe.

"In the beginning ..." "... בְּרָאשִׁית בָּרָא "

The opening passage of the Bible is familiar to us all. It starts
with the beginning of the universe. But what was happening *before* the beginning? Can we study, either theologically or scientifically, what there was before the beginning, if anything?
According to biblical tradition, what there was before the beginning is unknowable. This insight is based on the first letter of the
first word in the Bible. Tradition teaches that all aspects of the
Bible have significance, even the shapes of the letters. The first
letter of the first word of the Bible is the Hebrew letter bēth. It is
on this seemingly irrelevant fact that the sages based their understanding that any knowledge of what preceded the beginning
is unattainable by investigation. Why? Because the shape of bēth

is such that it is closed on three sides and open only in the forward direction, similar in shape to a *C* but with its opening facing to the left, as Hebrew is written from right to left. The sages saw a parallel between the written form of this opening letter, the bēth, and the study of the universe. Because the Bible begins with a letter that is bounded on all sides except the forward, so the events that occur after "the beginning" are those that are accessible to human investigation. Similarly, those that preceded the beginning, that is the creation, are not open to investigation.

This concept that events prior to the Big Bang are outside the sphere of human inquiry, is an intimate part of the Judeo-Christian tradition. Centuries ago it was stated by Nahmanides in the introduction to his *Commentary on Genesis*, and more recently, it was restated by the Vatican during a conference of scientists and theologians called in 1981 to discuss cosmological theories of the early universe. The church was careful not to repeat the error it had made when Galileo exhibited evidence that the Sun was the center of the solar system![1]

Old Testament theology talks in the language of man, while current cosmology makes its statement in scientific terms. It, too, teaches that pre–Big Bang information is not within our reach. Events preceding the appearance of matter and space cannot be studied. At the earliest instant about which we can theorize, temperatures exceeded $10^{32}°C$ or 10 million billion billion times hotter than the temperature at the center of the Sun. No one knows what was present at that explosive instant but whatever was there was in an exotic state of madly rapid motion. The matter and space of this moment were so tightly packed, so dense, that the violent collisions among the particles of matter and those packets of energy we refer to as photons were continually shattering each other into and out of existence. Energy and matter were in a fluid interchange, just as Einstein's most basic statement of relativity implies: $E = mc^2$ and, equally true, $mc^2 = E$. At this early time, the E, that is the energy, of the photons was converting into the m, the mass, of mc^2, and equally rapidly, this very m of the mc^2 was reconverting to E.

This melee of random high-energy collisions precluded any possibility of order in the energy or particles present, order that would have contained information related to what preceded that instant. Without order, information cannot be transferred across a sequence of time, a temporal interface, separating the "before" from the "after." Although arrived at from separate directions, the conclusions of biblical tradition and cosmology are the same: We are dealing with a blank page as far as precreation information is concerned.

In the thirteenth century, Nahmanides quoted a commentary on Genesis written 600 years before him, noting that prior to the existence of the universe, time did not exist.[2] This was learned from the wording of Genesis 1:5, ". . . and there was evening and there was morning, day one." It is not stated "the first day." For the use of first would have implied an already existing series of days or a continuum of time when in fact on this "day one" there had been no prior time to this one day. Not a before and not an after. There was nothing to which one could relate this day. It stood alone as "day one." For all the remaining days in that week of Genesis, the ordinal terms, second, third, etc., are used and logically so. By day number two, and thereafter, a series of days had been established. Although it is difficult to comprehend, the creation of the physical universe brought with it a concurrent creation of time.

Commenting on the first chapter of Genesis, both Maimonides and Nahmanides reached a similar interesting conclusion: Prior to the creation of the universe, space did not exist.[3] The creation of the universe brought with it not only the time into which it flows, but also the space into which it expands. Thus the presence of the energy and matter of the universe not only causes the existence of time, but also of space.

Note the difference between this account of our cosmic origins that results from the revelation at Mt. Sinai and the myriad of pagan myths that told of our beginnings. We have here no mythic cows, no gaping voids, no primeval superstuff to shackle a deity to the limitations and frailties of material existence. Even Plato and Aristotle, amid the intellectual society of Greece, were not able to grasp the biblical concept of creation. Although both believed that there exist a god or gods that exercise power over the universe, for the Greeks these gods were incapable of creating matter. Their gods were bound by, and dependent on, the matter of the universe.

If space and time did not exist, how was the universe brought into existence? Some 500 years ago, the cabalists theorized that at the instant of creation, God, filling all eternity, contracted. Within that contraction, the universe expanded. To form the universe, God chose from the infinite realm of the Divine, ten dimensions or aspects and relegated them to be held within the universe. These dimensions are hinted at in the ten repetitions of the statements, "and God said . . ." used in the opening chapter of Genesis. The cabalists believed that only four of the ten dimensions are physically measurable within today's world. The other six contracted into submicroscopic dimensions during the six days of Genesis. Today we refer to these four measurable dimensions as length, width, height, and time, the criteria by which we can describe our every move.

To conceptualize a contraction of dimensions, visualize an ordinary pencil. At close range, it is seen to have length, width, height, perhaps color and perhaps surface roughness. As we move away, these dimensions gradually become less clear until only a thin colorless line is visible. We have contracted the original five dimensions of the pencil into one.

With an amazing congruity, particle physicists now talk of the String Theory, a unified description of our universe in ten dimensions.[4] Having ten dimensions allows a unified explanation of the fundamental forces that operate among particles of matter. These dimensions according to the physicists are the four that we know, length, width, height, and time, plus six others. These six are contracted into a size far too tiny ever to be observed even by the best of microscopes, but their existence can provide continuity to our most profound—and current—speculations of the underlying structure of the universe.

Let us now compare what tradition and theology have to say about the time *after* the bang that brought us into being.

In the Viewer's Guide to Carl Sagan's *Cosmos* television series, there is an emphatically definitive statement about our knowledge of the very early universe and the universe of today. To quote from the Viewer's Guide: "Without these advances [in astronomy that have revealed that most galaxies are receding from each other], we would not suspect that the universe is expanding, we would have no inkling that the universe probably evolved from a primordial state of high density." A primordial state of high density is that tiny spot within which all the energy and matter of the universe was once concentrated and from which this energy-matter continuum expanded to form the universe of today.

Steven Weinberg, in *The First Three Minutes*, made a similarly unequivocal statement. "Our knowledge of the expansion of the universe rests entirely on the fact that astronomers are able to measure the motion of a luminous body in a direction along the line of sight."[5] From this measure of motion, we have learned that almost all of the luminous bodies we call galaxies are receding from our galaxy. This recession does not mean that our galaxy, the Milky Way, is at the center of the universe, nor does it mean that it is not at the center. To conceptualize this uncertainty in our position, think of a loaf of dough with raisins distributed throughout its bulk. As the yeast causes the dough to expand, each raisin gets farther from its neighbors. Imagine yourself standing on any raisin within the loaf. Because the motion of the expansion is inertial (smooth or uniform) there will be no sensation that your raisin is moving. Every other raisin will be seen to move away from you. Now move to any other raisin. The sensation will be identical. Because the principle of relativity is at work, your new home will feel as if it is at rest and the other raisins are in motion. It is no different than the confusion over which train is moving, yours or the one next to yours. What you have learned from the observation of the motion of the other raisins is that your universe, your loaf, is expanding. You have not learned where you are within the loaf or where you are going.

In a similar manner, studying the motions of galaxies throughout the universe doesn't reveal where within the universe the Milky Way stands or how fast it is moving. What astronomers have measured is the relative motion of the galaxies, and they have interpreted this information to indicate an expanding universe. With this knowledge, science, as Sagan and Weinberg so definitively stated, has staked its claim to the discovery of the expansion of our universe. According to these observations, if the

Big Bang theory is correct, then biblical tradition will have to accommodate itself to include this fact of expansion.

The statements of Weinberg and Sagan that I have quoted are almost, but not quite, correct. That is to say, the universe is expanding. However, the good news for those who hold by the Judeo-Christian tradition is that mankind's awareness of the expansion is not quite so new as Weinberg and Sagan would have us believe. You see, had they studied the history of cosmological thought contained within texts of their own tradition, they might have modified the "no inkling" statement to read something like: Without the recent advances in astronomy or knowledge contained in Nahmanides's *Commentary on Genesis*, we would have no inkling that our universe is an expanding universe. The fact, which is startling if one considers its implications, is that both radioastronomers and biblical scholars have discussed the events of the Big Bang and have done so in terms that are uncannily similar. We know the sources of the cosmologists' data, but we can only speculate on the sources used by the sages.

Theories of the very early universe proposed by present-day astrophysicists and the revelations brought by the sages both deal with conditions at about the same time, when everything was so tightly compressed that matter as we know it today could not possibly exist. I will deal with the beginning in two stages: the material present at the very beginning and then the events of this earliest period of our ancestry.

In the beginning God created . . .

(Gen. 1:1).

The Talmud[6] states that there are two incidents in the Bible of such profundity that it is forbidden for a teacher to expound on them with more than one or two selected students. Even then, the teacher is to acquaint these students only with the "section headings." The incidents are the account of the beginning (Gen. 1) and Ezekiel's vision of the heavenly chariot (Ezek. 1). This Talmudic statement relates to a time when all the knowledge revealed to Moses at Mt. Sinai was still remembered. Much of that knowledge is long since lost. Sages thereafter could only find hints of it by searching out the subtle meanings of the biblical text. Based on these hints, they formed concepts of a space-time continuum that comprises the universe.

Tradition has taught us that at the instant of creation time and space came into being. Prior to this instant, none of the ever-so-common lengths and widths and heights or the time within which we so casually live out our lives existed. Such a condition almost seems to be an impossibility. We cannot visualize a total space-lessness and timelessness.

With the creation of the universe, space expanded and time extended as the very existence of the universe pushed back the borders of space and time. Space as we know it ends at the edge of the universe [there is no outside to our universe]. We may close the door to a room but we are well aware that the space on the other side of the door continues to exist. We are familiar only with the sort of boundary that has something on the other side.

The same is true of time. The extension of this fourth dimension, time, is measured from the instant of the creation and extends only to the present moment. As the universe continues to exist, time is extended to encompass its existence. Tomorrow may be written on a calendar but it comes into existence only when, in the time dimension, the universe flows into tomorrow.

The intimate relation among the space, time, and material existence of our universe was to remain a speculative observation until it was quantified by Einstein in his theory of relativity. With uncanny but reassuring similarity, the theory of relativity deals with perceptions of the same three quantities discussed by traditional sources: space, time, and mass.

The creation of the heavens and the Earth from absolute nothing is at the root of biblical faith. "He who does not believe in this and thinks that the heavens and the earth have existed forever denies the essential and basic principle of [biblical] religion,"[7] wrote Maimonides. The Hebrew word used for creation, barah, is the only word in the Hebrew language that means the creation of something from nothing. Biblically it is applicable only to the actions of God. It is the second word of the Bible.

As theoretical physicists attempt to describe the very early universe, say 10^{-43} (that's 0.0000000000000000000000000000000 0000000000001) seconds after the creation, they have also been confronted with the image of a universe devoid of material existence. As the theories of the early universe reach back to the beginning, they describe a condition in which all the matter is pressed into a space of zero size and infinite density. Infinity cannot be dealt with quantitatively and so cosmologists cannot describe the conditions of our absolute origin in real terms. Only by working in a dimension known as imaginary time, a concept that does not translate into a dimension of the world in which we live, can the very instant of the beginning be described mathematically. But if we relate to real-world dimensions, that zero point of time, the beginning, is beyond the grasp of mathematics and physics.

Biblical tradition does not envision the early universe as flowing from an infinitely small point. Biblical tradition describes the start of the universe as occurring in a tiny but finite speck of space, about the size of a grain of mustard.

For conditions just after time zero, there is agreement between science and theology. Both propose that if matter was present, it was present in minuscule amounts relative to the quantity of energy present. In characterizing the start of the universe as being essentially massless, Alan Guth of M.I.T. has referred to the universe as possibly being "the ultimate free lunch."[8]

When considering the events and even the nature of matter preceding the creation, as well as the creation itself, we find literal agreement between science and theology. This agreement that the universe formed from an initially massless state should not be discounted as trivial. If the theological description of the universe were based on a simple understanding of nature, then there would be no agreement. Because creation ex nihilo cannot be learned from simple observations of nature, both Plato and Aristotle believed that the matter of the universe must be eternal.[9] A towering tree may form from a small seed or a human from a tiny ovum, but the seed or ovum need a continuous flow of nutrients. The cosmic analogy of this, as envisioned by the Greek philosophers, was a need for a primeval substance predating the universe from which the universe was formed. In contrast, the founders of the Judeo-Christian theology considered that "nothing" was quite sufficient to provide the start of our universe.

Part of the study of the matter or lack of it at this very early time is based on the search for the most basic building blocks of matter. The elementary particles of our universe are far too small to be observed by even the most sensitive of microscopes. To learn about the properties of elementary particles we must study the effects that they have when interacting with matter. This is reminiscent of the study by theologians of the attributes of God. These attributes are not learned by direct observation, but only by evaluating the effects of interactions between God and the universe.

The start of the universe, according to biblical tradition, is described by Nahmanides in his *Commentary on Genesis*. Nahmanides's amazingly perceptive description of the early moments of the universe presents a scenario totally unlike normal, or even abnormal, human experiences. Considering the lack of scientific knowledge at the time that Nahmanides lived, he must have had either extraordinary gumption or absolute faith to have taught such notions. In fact, he explicitly states in the introduction to his biblical commentary that his description will be incomprehensible to readers who have not "received hidden wisdom." "I give proper counsel to all who look into this book not to seek explanations of the hints I write concerning hidden matters of the Torah. For I make surely known that the reader will not grasp my words by reasoning."

Nahmanides bases his warning on a verse in Ecclesiastes (9:11), "I returned and I saw under the sun, that not to the swift is the race and not to the mighty the battle and also not to the wise, bread and also not to those who understand, wealth and also not to those with knowledge, grace, for time and happenstance occur to all."

Note the phrase, "and also, not to those with knowledge, grace." There seems to be no logic to the juxtaposition of "knowledge" and "grace." In Hebrew, the word for "grace" is a two-letter word, *hn*. These two letters are the initial letters of the Hebrew words *hochmah nistarah*, which translates literally as "hidden wisdom." Study alone is not sufficient to derive all the hidden truths of the Bible. There has to be revelation as well. Learning even part of these truths without the help of "grace" requires, according to Nahmanides, a deep understanding of the natural sciences of the world.

Nahmanides's account of the first seconds of the universe reads like this: At the briefest instant following creation all the matter ✓ of the universe was concentrated in a very small place, no larger than a grain of mustard. The matter at this time was so thin, so intangible, that it did not have real substance. It did have, however, a potential to gain substance and form and to become tangible matter. From the initial concentration of this intangible substance in its minute location, the substance expanded, expanding the universe as it did so. As the expansion progressed, a change in the substance occurred. This initially thin noncorporeal substance took on the tangible aspects of matter as we know it. From this initial act of creation, from this ethereally thin pseudosubstance, everything that has existed, or will ever exist, was, is, and will be formed.[10]

Nahmanides's reference to a grain of mustard is the traditional way of saying, "in the language of man," the tiniest imaginable speck of space. Nahmanides taught that at the beginning, all that is on and within the Earth and all the heavens, in fact all the universe, was somehow packed, compressed, squeezed into this speck of space, the size of a mustard grain.

Before we convince ourselves that Nahmanides had let his imagination run wild, let us listen to the description that current cosmologists have for that same early time in our evolution.

The present universe, according to current cosmological understanding, is the result of a Big Bang, a massive expansion from a single point. While the conditions that existed prior to the appearance of energy and matter are not known, we can attempt to describe them at the briefest instant following the beginning, at about 10^{-43} seconds after the start. That time reads as one 10 millionth millionth millionth millionth millionth millionth millionth of a second. The universe was then the size of a speck of dust. It would have taken a microscope to study it. Now, 15 billion years later, even telescopes are not powerful enough to reach its limits.[11]

At this early time, all matter was concentrated into the one minuscule core location. The temperature was $10^{32°}$ K (100 million million million million million degrees Kelvin). For comparison, the temperature at the center of the Sun is about 15 million degrees Kelvin. The surface of the Sun is a mere 5,800°K.

Physics and mathematics, as we know them today, cannot deal with times earlier than 10^{-43} seconds after the beginning. Prior to that time, the temperatures and densities of matter exceeded those that can be described by the laws of nature as we now understand them. Because of this, cosmological theory cannot handle the actual time-zero beginning of the universe in terms that relate to dimensions experienced by humans.

As the study of events following the Big Bang is extended mathematically to earlier times, the size of the universe shrinks toward zero and, inversely, the temperature and density increase toward infinity. The actual instant of the beginning envisions, for physicists, a moment when an infinitely small point of space was packed with matter squeezed to an infinitely high density. This condition of infinities is referred to as a singularity, and singularities cannot be treated by conventional mathematics. Only by transferring the numbers of the equations into what are called "imaginary units" can we avoid the singularity that arises for cosmology "in the beginning." But in real terms, in units able to be sensed by humans, the singularity remains. The conditions of time zero elude us. In other words, although there is a theoretical solution in the world of physics to this problem of the beginning, in terms that are perceivable by humans, there is no solution.

In very early times, matter was not matter as we know it. The high pressure and temperature in this core had reduced all matter to its form of pure energy. The concept of matter, even of the tiny theoretical fundamental particles called quarks, has no meaning for the temperature, pressure, and spacial dimensions that are speculated to have existed at this very early time. There was exquisitely hot energy and very little else. Within the initial core location, an explosion or inflation occurred that forced the energy-matter out in all directions. The cause of this inflation is not clear. Some scientists posit that mutual repulsion among all that was present occurred, something akin to a force of antigravity. The term inflation is used deliberately. It implies that the forces that pushed back the boundaries of space to the size of a grapefruit came from within. There was no without. There was and is the universe and the space it occupies. That was and is the totality of all physical existence.

Concurrent with the expansion there was a lowering of pressure and temperature. At these more moderate, expanded conditions (a mere billion billion billion degrees Kelvin), energy could now condense into the tiniest of particles, the theoretical quarks and the known electrons. This took place in accord with Einstein's law, $E = mc^2$, which states that energy and mass are actually different states of a single energy-matter continuum, just as water, steam, and ice are all composed of a single entity, H_2O. Energy is matter in its intangible form; matter is energy in its tangible form.

As this expansion progressed out, away from the core, pressures and temperatures fell. Conditions became less harsh. The transition of energy to the more substantive forms of tangible matter continued. The material universe as we understand it came into being. The entire process is referred to as the Big Bang. Thus far, the cosmological description of the early universe.

This process is so widely accepted by researchers currently active in cosmology that it is referred to as the Standard Model of the formation of our universe. Its adherents include A. G. W. Cameron, Frank Press, Raymond Siever, and Steven Weinberg, but not Albert Einstein in his initial formulation of the model of our universe, as we shall see later.

The parallel between the opinion of present-day cosmological theory and the biblical tradition that predates it by over a thousand years is striking, almost unnerving. In view of the radical departure from our conception of reality, it is not surprising that even a Nobel laureate, such as Steven Weinberg, would think that unaided human perception is incapable of having an inkling of these occurrences that marked the evolution of our early universe.

The question we face is, From what source came the aid in perception? I think that we can rely on our understanding of history. Ancient biblical scholars did not have the aid of radio-astronomy or spectroscopy. So how did they have the insight, a thousand years ago, to form an account of the Big Bang so strikingly similar to our modern theories? How could these early teachers have known of our origins within a speck of space, of the expansion of space that has led to our universe, of the transition from an ethereal nonsubstance to tangible matter, and, even more precisely, that this transition from formlessness to matter with form accompanied the expansion of the universe?

When the writers of the *Cosmos* series claimed that without the modern equipment available to those involved in cosmic research we would not suspect that the universe is expanding, that we would have no inkling that the universe probably expanded from a primordial state of high density, that is, we would not have discovered the phenomena of the Big Bang—they were, of course, correct. Discovering the phenomena related to the original Big Bang required sophisticated radio and optical telescopes and all the technology related to high-energy particle accelerators. These became available only in the last 50 years or so. Data had to be gathered and correlated and inferences had to be drawn. Nahmanides and Maimonides were not in the business of discovering. For them, all could be derived from the revelation associated with the Bible. As Nahmanides stated, "What other source would be used?"[12]

Consider the position of a teacher of natural sciences a thousand years ago suggesting that in the beginning all that is now our universe was contained within a single location no larger than a grain of mustard. A skeptic places before the teacher a glass of water and asks the teacher to compress it to half its size. Impossible in human experience. How much less possible is the compression of all the contents of the Earth and then of the universe into a space the size of a grain of mustard? The response to the skeptic is not one of proof. It must be one of faith; faith in the accuracy of revelation even when it precedes the advances of science that eventually come to confirm it.

Revelation, at least as we have it today, did not provide details. At normal temperatures and pressure, matter is arranged in molecules. As pressures increase, the molecular structure is destroyed and individual atoms remain. Increasing the pressure even more destroys atomic structure until only atomic nuclei and free electrons exist. Finally, even the nuclei are pressed so tightly together that they break. When the compression finally results in temperatures that exceed the rest energy of these particles, that is, when the E is greater than the corresponding mc^2, the particles freely transform from their mass form into energy.

Mankind, formed from the primeval energy of the Big Bang, can discover the details from physics just as he could receive them from biblical revelation.

Notes

1. For a description of the proceedings of this conference see: Hawking, *A Brief History of Time*, pp. 115–116.
2. Nahmanides, *Commentary on the Torah*, Genesis 1:4,5.
3. Maimonides, *The Guide for the Perplexed*, part 2, chapter 13.
4. For a discussion of String Theory see: Crease and Mann, *The Second Creation*.
5. Weinberg, *The First Three Minutes*, p. 12.
6. *Babylonian Talmud*, Section Haggigah 2:1.
7. Maimonides, *The Guide for the Perplexed*, part 2, chapter 27.
8. Gore, "The once and future universe," *National Geographic* 163 (1983):741.
9. Cohen and Drubkin, *A Source Book in Greek Science*.
10. Nahmanides, *Commentary on the Torah*, Genesis 1:1.
11. For two very readable elaborations of this theory see: Hawking, *A Brief History of Time*, and Weinberg, *The First Three Minutes*.
12. Nahmanides, *Commentary on the Torah*, Introduction.

CHAPTER 4

Beginnings

■■■■■■■■■■■■■■■■■■■■■■■

The account of the beginning [Genesis 1] is natural science but so profound that it is cloaked in parables.

—Maimonides

T hree theories of origin of the matter found in today's universe are currently proposed. The most widely accepted theory has attained the title of the Standard Model. In essence, this theory describes the processes of the Big Bang. The other two theories are of an oscillating universe and of a universe in a steady state.

The oscillating universe theory is a modification of the Standard Model. Philosophically this theory, which is similar to Plato's idea of an eternal but changing universe, solves all of an agnostic's problems with the existence of the universe. It proposes that the universe has always existed. Following a Big Bang, energy and matter move out from the central location of the Big Bang propelled by the force of the explosion. The pull of gravity among the masses of gases, stars, and galaxies in the expanding universe is proposed to be so strong that gradually the rate of expansion slows. Over tens of billions of years, this gravitational braking of the expansion succeeds in stopping the expansion. The unending inward pull of gravity among all the mass in the universe gradually causes an inward flow of the matter of the universe toward the former center, the location of the previous Big Bang. Tens of billions of years pass. The continual inward flow has now compressed all matter again into the center of the universe. The high pressures and temperatures in this core convert the stuff of the universe from its matter form to its energy form. A new Big Bang occurs and another expansion-braking-contraction cycle starts— *ad infinitum.* No creator is needed.

The steady-state theory, formulated by H. Bondi and T. Gold

in the 1940s, is quite similar to Aristotle's view of the universe. It proposes that the universe has always existed and will always continue to exist in a condition similar to its present condition. As galaxies move apart and as stars consume their nuclear fuel, cool, and die out, new hydrogen is created from nothing to fill the resultant spaces. From this hydrogen comes the substance of which new stars are made.[1] The theory does not mention the source or force creating the new hydrogen. But some matter has always been present. It is a totally hypothetical concept. The creation of new hydrogen has never been observed.

Although some scientists may retain their belief in the steady-state theory of the universe, two recent discoveries strongly indicate that some form of Big Bang followed by an expansion is part of our cosmic heritage. A steady-state universe with continuous production of hydrogen and no Big Bang is inconsistent with recent cosmological discoveries.

The first discovery is that of the Doppler shift in the light emitted by galaxies. In 1842 J. C. Doppler, a professor of mathematics in Prague, described a phenomenon that was to be known by his name, the Doppler effect. Briefly stated, all transmissions of energy having wavelike characteristics (such as sound and light) that are emitted from a moving source have an apparent shift in the frequency of these emitted waves relative to a stationary receiver. This change in frequency is referred to as the Doppler effect. The extent of the shift in frequency depends on the speed and direction, that is, the velocity, of the emitter relative to the receiver. We can all experience the Doppler effect for sound waves by standing by the side of a road and listening to a car horn blown continuously as the car passes. In this case, the emitter is the car's moving horn and the receiver is our ears and brain remaining stationary by the side of the road. The pitch of the horn is heard to rise as the car moves toward us. When the car passes and moves away from us the pitch lowers. As the car approaches us, the sound waves actually compress. As each successive sound wave is emitted, the car is a bit closer to us than when the previous wave was emitted a fraction of a second before. Because the car is closer, the new wave travels a slightly shorter distance than the previous wave and so reaches us in a bit less time. This shorter travel time in effect increases the frequency (the number of waves per second) at which the waves reach our ears. The increased frequency is heard as a higher pitch.

As the car moves away, the opposite effect occurs. The wavelength then stretches out or elongates. The sound becomes lower in frequency or pitch.

The same phenomenon, although usually not detectable by unaided human senses, is true for light waves. A spectrometer shows that light emitted from an object moving toward us is shifted toward the blue end of the color spectrum. Blue light has a relatively short wavelength and hence a relatively high frequency. Light emitted from an object moving away from us shifts toward the red, that is, toward lower frequencies and longer wavelengths. The amount of the shift is a measure of the velocity of the object relative to the observer (see Figure 3).

In the early 1800s, Joseph von Fraunhofer, an optician in Munich, noticed that when a thin sliver of sunlight passed through

Figure 3. *The effect of relative motion on the frequency or pitch of sound and on the frequency or color of light.*

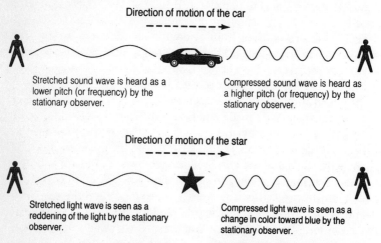

Direction of motion of the car

Stretched sound wave is heard as a lower pitch (or frequency) by the stationary observer.

Compressed sound wave is heard as a higher pitch (or frequency) by the stationary observer.

Direction of motion of the star

Stretched light wave is seen as a reddening of the light by the stationary observer.

Compressed light wave is seen as a change in color toward blue by the stationary observer.

Note that, due to the differences in the motion of the car or the star relative to each observer, the noise from the same horn is heard as two different pitches and the light from the same star is seen as two different colors.

a prism, the spectrum produced by the prism was crossed by hundreds of black lines. These lines were also found in sunlight reflected from the moon and in the light of stars. They are the result of selective absorption of specific light frequencies as the light travels from the star surface through the cooler gases in the star's outer atmosphere.

Careful comparison of these absorption spectra among the light of different stars shows that the locations of the lines are often slightly displaced from star to star. Sir William Huggins, in 1868, realized that this displacement was a Doppler effect resulting from the motion of the particular star relative to the Earth. The measurement of this Doppler effect produces an accurate measure of the velocity of the star relative to the Earth.

In the early 1900s, Edwin Hubble assisted by Milton Humason, a spectroscopist, spent three decades measuring distances to galaxies and the shift in the spectrum of light emitted by the stars of these galaxies. They discovered that most of the light showed a shift toward the red. That meant that the galaxies were moving away from us. Equally interesting, they found that the amount of the red shift was approximately proportional to the distance of the galaxy. They interpreted this observation to mean that the galaxies were moving away from us at velocities that were proportional to their present distance from us.

These results and those of astronomers who continued their work in the decades that followed revealed a consistent distribution of relative velocities among most of the galaxies. Almost all are moving away from a common universal center at speeds proportional to their distance from that center. That is, the farther away a galaxy is, the faster it is moving away. This pattern is logical if all matter started moving out from a single, common, central location at the same instant. That matter that received the highest initial speed still travels the fastest and so has moved out the farthest from what was the center. The matter that received a lesser speed at the start moved out more slowly and so has traveled less far from the center throughout the billions of years that have followed the big explosion, the instant of the Big Bang. Because we are also within a moving galaxy, we cannot locate where this universal center is any more than someone on a raisin within the loaf of rising dough could tell where the center of the loaf was. All we can do is measure the galactic velocities relative to our own motion through space.

It is the Doppler effect that Weinberg was referring to when he stated, "Our knowledge of the expansion of the universe rests entirely on the fact that astronomers are able to measure the motion of a luminous body in a direction along the line of this flight."[2] As we learned, the sages quoted by Nahmanides discussed this "recently discovered" universal expansion more than 2000 years ago! But they had only revelation, not Doppler, on which to rest their case.

The second recent cosmological discovery indicating that something akin to a beginning is part of our heritage is that of the existence of an isotropic radiation background, that is, a radiation background that is constant and equal in all directions of the universe. This was discovered in 1964 by Arno Penzias and Robert Wilson at the Bell Telephone Laboratory located at Crawford Hill, New Jersey.[3]

To understand the significance of the discovery of this uniform radiation, we first need to explore a bit of radiation theory and terminology. We all experience *thermal radiation*—the Sun's warmth or the warmth we feel in the evening radiating from a stone wall that has absorbed sunlight during the late afternoon. This type of radiation ceases only when the object has cooled to absolute zero. Absolute zero is reached at $0°K$ or $-273°C$. At this very chilly temperature, all thermal motion of the electrons has stopped and so there is no further thermal radiation. It is absolutely zero!

Radiation is often discussed with reference to the temperature to which an opaque, colorless (or black) object must be heated to emit the identical radiant energy that is coming from the place or object being studied. It is remarkable that this radiation is independent of the material from which the object is made. It depends only on the temperature of the "black body." In fact, an ideal black body is not really an object at all. It is the radiation measured in the cavity of a box having opaque walls.

Although at absolute zero all thermal motion ceases, at all temperatures above absolute zero, thermal radiation is constantly being emitted. Humans and most higher animals are quite sensitive to radiations associated with black body temperatures of several thousand degrees Kelvin. We call this radiation *light*, and we measure its wavelength or frequency with our eyes as the different colors of the spectrum. Nerves in our skin are receptors for radiation associated with temperatures of hundreds of degrees Kelvin. This we feel as infrared heat. Radio waves are examples of radiation from very low temperature bodies. So are microwaves. At radiant temperatures much higher than those seen as light, the radiation enters the region of X rays and gamma rays. This radiation is so powerful that it can penetrate our bodies. The general relationship among the different energies of radiation is that as the energy gets higher, the frequency also gets higher and the wavelength gets shorter. The higher the radiation energy, the higher the temperature of the black body that will emit this radiation. Regardless of the particular energy, all radiation is carried by photons. Table 5, in the appendix of this book, shows the common names for electromagnetic radiation in relation to their wavelengths, frequencies, and black body temperatures.

Penzias and Wilson used a horn antenna, originally designed for communication via the *Echo* satellite, to study the radiation characteristics of space. This special antenna received radiation only from the small segment of the sky toward which it was pointed. Penzias and Wilson observed that regardless of the direction the antenna was pointed toward the heavens, considerable radiation energy was always present at a wavelength of 7.35 centimeters, the microwave region of the energy spectrum. It was as if the universe was permeated with this microwave background. This wavelength corresponds to radiation that would be emitted by a black body heated to 3.5 degrees above absolute zero, or 3.5°K.

The discovery of the 3.5°K radiation background created a sensation in the community of astrophysicists. It confirmed a crucial prediction of the Big Bang theory. If the estimates of the conditions during the first few minutes following the Big Bang are correct, there should be a black body radiation background of about 5°K evenly distributed throughout the entire universe. Its presence is the residue of the intense, high-temperature radiation that must have existed at the start of the universe. Without the shatteringly disruptive effects of an intense radiation field, the particles present in the early universe would have been literally squashed together. They would have been fused into heavy elements by the immense pressures that existed during the early phases of our universal expansion. The result would be a universe composed predominantly of heavy elements. But this would be inconsistent with observations showing that approximately 75 percent (by weight) of the matter in the universe is hydrogen and the remaining 25 percent is predominantly helium. Greater than 99 percent of the mass of the universe is in the form of hydrogen and helium, the two lightest elements of the 92 elements occurring naturally in the universe.

Hydrogen consists of a single proton in its nucleus surrounded by a single electron. As such, it is the lightest element. Helium has two protons and two neutrons in its nucleus; it is the second lightest element. Elements heavier than helium might have formed during the first moments following the Big Bang. This could have been accomplished by fusing more protons and neutrons into helium nuclei. If they did form momentarily, they were literally smashed apart by the intense radiation present at that time. Even hydrogen and helium existed only as atomic nuclei. Their electrons were stripped off by the radiation. The formation of the heavy elements that we know today (for example, carbon with its six protons and six neutrons in its nucleus and iron with its 26 protons and 30 neutrons) took billions of years. It took (and still takes) place only within massive stars where the already existing hydrogen and helium are pressed and fused together to form the elements of life.

But what is the connection between today's universal radiation background associated with a chilling 3.5°K above absolute zero and the radiation at the time of the Big Bang, a radiation measured as billions of degrees? A clue to the answer is found in the cooling coils of your refrigerator.

As the supercompressed stuff of the early universe expanded, the temperature fell. The temperature had to fall because cooling is an inherent characteristic of an expansion. It is, in fact, the expansion of a compressed gas, often Freon, that cools your refrigerator. As the universe expanded from being packed into the size of a mustard grain to the immensity of today's volume, the initial unimaginably hot 10^{32}°K temperature cooled to a chilling

3.5°K. That is 269 degrees below the freezing point of water! The expansion of the universe stretched the wavelengths of the initial highly energetic photons from billionths of a centimeter to the 7.35-centimeter wavelength discovered so recently by Penzias and Wilson. At this wavelength, the temperature of photons, once so hot, is now a mere 3.5°K. This 3.5°K radiation is perhaps the only measurable legacy remaining from the immensely hot beginning of our universe. It is a sort of fossil in photons.

As a result of these two discoveries, the Doppler effect showing an apparently expanding universe and the isotropic cosmic radiation at 3.5°K resulting from this expansion, current cosmologists accept the Big Bang as the Standard Model. It is as much a part of our individual histories as are our family ancestors.

The Big Bang, however, does not necessarily require a Creator. The theory that describes the universe as being in a state of perpetual oscillation still might rescue the skeptic from the need for a Beginner that might be associated with this beginning. Recent cosmological observations, however, place the possibility of an oscillating universe in serious doubt.

A universe in oscillation is seen to be in an unending state of cyclic motion: expansion, gradual slowing of this expansion, contraction, implosion, and explosion. A fundamental principle of motion, that of inertia, was first deduced by Isaac Newton. It states that an object at rest will remain at rest until acted on by a force. If the force gives it motion, then the object will continue to move in a straight line until other forces change that direction and/or speed of motion. The measure of this tendency of an object to remain in linear motion is referred to as the object's momentum.

At the instant of the Big Bang, the substance of the universe received a powerful push. The momentum that this push instilled in that substance is what has kept us flying through space for the last 15 billion years. It was the single source of all the motion of the universe, and all subsequent motion derives from it. But the expansion of the universe does not proceed freely. An inherent characteristic of matter is the mutual attraction it has toward other matter, the force of gravity. What produces gravity, how it travels through space; these are questions still unanswered by science. But the force of gravitational attraction between objects can be measured. It is proportional to the product of the masses of the objects and inversely proportional to the square of the distance separating the objects. It appears to travel with the speed of light.

If the universe is to be in a state of perpetual oscillation, there must be sufficient matter in the universe to cause the gravitational attraction to eventually overpower the momentum of expansion. If there is not enough mass to cause this, the universe will expand forever or at least until acted on by a force that we have not yet observed in nature.

The thrust of the Big Bang was so strong that the galaxies are still flying apart, out from the common center. But the gravitational pull inherent in the mass and energy (it is now understood that gravity is generated by all forms of energy and not only by the form of energy we refer to as matter) of the universe is slowing their outward flight. If there is enough mass present, the outward flight will gradually, over billions of years, be slowed, stop, and retract into the center just as a ball thrown up from the Earth will be pulled back to the Earth. The result could be another Big Bang.

The amount of mass required to cause this retraction, that is, the average density of cosmic matter, is referred to as the critical density.[4] The predicted value of the critical density is only an average of a few hydrogen atoms per cubic meter of space. If you realize that in one cubic meter of water there are about 10^{29} or 100 billion billion billion hydrogen atoms, it becomes clear how tiny an average cosmic density is needed. But the present universe is immense and even at two or three atoms per cubic meter we are dealing with vast numbers of atoms and hence vast amounts of mass.

Current estimates of the cosmic density indicate that in the entire universe only 10 to 20 percent of the matter required to cause the eventual contraction of the universe exists. There is simply not enough matter to produce the gravitational force needed to stop the outward flight of the myriad of stars and galaxies of the universe. This conclusion is consistent with analyses of Doppler shift data, which also predict no end for the expansion of the universe. Comparison between the Doppler shifts in light received from galaxies and the distance that these galaxies are from the Earth indicates that their rate of flight is not slowing down sufficiently to cause them eventually to stop their outward expansion. What this means is that the expansion of the universe will continue. Eternally. There was only one Big Bang.

A Big Bang followed by an unending expansion of the universe tells us that there was a beginning and that, at the minimum, there is a place for a Beginner.

There still remains a residue of doubt concerning the total mass of the universe. The pattern of galaxies throughout the universe suggests to some theoreticians that there is a gravitational force influencing the motion of the galaxies. The visible matter of the universe is not sufficient to account for this force, so they conjecture that in addition to matter actually observable in the cosmos, there exist large amounts of invisible matter.

A prime candidate for this invisible matter is the large quantity of neutrinos believed to have been produced during the first seconds of the universe's existence. Neutrinos are tiny, chargeless particles that rarely interact with other forms of matter. Millions have just passed through the book you are reading and also through you. The question is, Do neutrinos have a rest mass? If they do, this would add considerably toward the total mass of the universe and possibly shift the gravitational balance toward the eventual collapse of the universe.

An exceptional event occurred on the night of February 23, 1987. Light from an exploding star, a supernova, reached the Earth. It had taken 170,000 years for the light, streaking through space, to arrive. At the time of that supernova, the forefathers of Neanderthal man were making axes from stone. Had they looked at the heavens they would have noticed nothing unusual. Although the explosion occurred then, the location was 170,000 light-years away. The light of this supernova, called 1987A, would reach Earth only 170,000 years later.

Supernovas have been believed to be producers of large numbers of neutrinos. Sure enough, detectors in Japan and the United States recorded bursts of these elusive particles at approximately the same time as the arrival of the supernova's light. After 170,000 years (Earth's time frame) of travel, the photons (those packets of energy with zero rest mass that we call light) and the neutrinos arrived at the same time. To arrive at the same time means that both had traveled at the same speed, c, the speed of light. Einstein's law of relativity shows that only particles with zero rest mass can travel at c. As John Bahcall of the Institute for Advanced Studies at Princeton said, because the neutrinos arrived at approximately the same time as the light, their rest mass must be vanishingly small, "the neutrinos can't contribute noticeably to the problem [of total universal mass]."[5] And so current estimates of cosmic mass indicate that the universe will *not* collapse.

There is one more clue we have in this search for a solution to the expansion of our universe. That relates to the entropy of the universe. Entropy is a measure of the disorder, the chaos, in a given system. I plan to discuss the mechanics of entropy in detail when we delve into the wonders of getting order out of chaos. For the present discussion, it is sufficient for us to realize that when left unattended, everything—whether inert or living, complex or simple—tends toward ever-increasing entropy, ever-increasing disorder.

The implications of increasing entropy in terms of cosmic thermodynamics are considerable. They have been dramatically described by Steven Weinberg in his landmark book, *The First Three Minutes*. In each cycle of expansion and contraction in an oscillating universe, the entropy must increase. Such is the nature of expanding and contracting fluids. This is why your refrigerator needs a motor. It takes the extra source of energy from the motor to put order back into the Freon (that is, to recompress it) after it has expanded and cooled the cooling coils of the refrigerator. If the universe is indeed in an unending series of cycles, then in each cycle the entropy must have increased. This increase would appear as an increase in the number of photons relative to the number of particles with a rest mass. It is these same particles that make up the matter of the universe.

Now if the law of entropy, which is always true on Earth, is cosmically applicable (and recall that all leading physicists assume the laws of physics that we observe on Earth are applicable throughout the universe; without this assumption there is no basis for any calculations of cosmology), then in the next expansion and contraction cycle of the universe, there will be more photons relative to particles than we have at present.

But there's a paradox here: If we are in an eternally oscillating universe, then there was no beginning. Time extends back to an infinite past. To reach the present, the "now" of our existence, from an infinite past would require an infinite number of cycles of expansion and contraction. That infinite number of cycles would, according to thermodynamics, have raised the ratio of photons to nuclear particles to infinity. Because the number of photons is finite, for the ratio to equal infinity, the number of particles must equal zero. Zero particles means that there would be no material universe. Photons would be the only components. But our very existence attests that this is not the case. There is a material universe and we are part of it. Based on this, Weinberg was led to ponder that "it is hard to see how the universe could have previously experienced an infinite number of cycles."[6] There must have been a beginning.

The evidence presented by today's leading cosmologists indicates that the present expansion of the universe resulted from a Big Bang, that the present expansion is not self-limited, and that it will continue. The universe is not in a state of cyclic oscillation, of expansion and contraction. There was only one Big Bang at which time our universe was created.

It is written in Psalms 148:4–6, "Praise Him heavens of heavens and the waters that are above the heavens. Let them praise the name of God, for He commanded and they were created. And He established them for ever and ever; He gave a law and it will not be transgressed."

Based on these verses, the sages have said that the universe can exist forever. But if its destruction is to come, this will not be as a natural act. It will be as a special force from God just as was its creation.

The heavens are eternal. This is true according to the Bible's description of our universe and according to the best estimate that science has for the situation.

AN IMPORTANT LESSON FROM HISTORY

In a sense astrophysics is part of the supersearch for a Creator. As you might expect, astrophysics and biblical scholarship are on the same team, only not all those involved in the search realize this. And that is why both theologian and scientist must beware of subjectively filtering the data gathered in this search. At times the temptation is to use only the agreeable information and neglect those data that seem to contradict a preconceived notion of the truth.

Johannes Kepler succeeded in deducing that the planets revolve around the Sun in elliptical orbits and not circles, because he had the courage to use data that contradicted the generally held concept that all orbits were purely circular.

Newton stated that an object in motion will remain in motion until acted on by an outside force. This was blatantly ridiculous. Not only did it contradict Aristotle's notion that the natural state of all objects is rest, but it also contradicted human experience. Rolling balls always eventually stop rolling.

Think of the professional and social pressure these scientists must have felt to adhere to the prevailing opinions. The psychological strain placed on people with unpopular theories is immense. Consider the following: In 1906 Ludwig Boltzmann, one of the founders of statistical mechanics, committed suicide. One of the causes of this tragic event was the intense philosophical opposition to his work, which now forms an integral part of physics. Not every scientist can withstand the force of that which is accepted as the current truth.

During the accumulation of the data that led to the widespread acceptance of the Standard Model of the universe there occurred an interesting lesson in nonobjectivity by the greatest of scientists. Albert Einstein completed his famous and complex general theory of relativity in 1915. Almost immediately he attempted to solve the relativistic equations to gain a description of the space-time physics of the entire universe. At that time, the current cosmology considered the universe to be isotropic and without expansion. Doppler shifts had not yet been measured in light emitted by galaxies distant from the Milky Way. Einstein's solutions of his general theory, published in 1917, correctly revealed a universe with expansion but such a concept was not in vogue at the time.[7] Relying on the then-current cosmology, Einstein introduced "a cosmological constant" into his equations. In so doing, he forced his relativistic equations to describe a universe without expansion. Years later, Einstein considered this one of the worst errors of his professional life.

The cosmological constant was no more than what a college freshman would call a "fudge factor," a totally subjective modification of the objective solution he was seeking. It forced his equations to give the desired answer. Five years later the mathematician Alexander Friedmann resolved Einstein's equations omitting the fudge factor. The solutions Friedmann obtained revealed a universe in expansion. His results showed that if the amount of mass present in the universe was less than a certain critical amount, the universe would expand forever. If more than the critical mass was present, the expansion of the universe would eventually cease and then contract. As we have seen, all the current data point to an ever-expanding universe, as per Nahmanides's presentation of biblical tradition of a thousand years ago.

Notes

1. Bondi, *Cosmology.*
2. Weinberg, *The First Three Minutes.*
3. For an excellent description of their contribution to cosmology see: Weinberg, *The First Three Minutes.*
4. Discussions of critical density may be found in the books on cosmology cited in the Bibliography and in these journal articles: Gore, "The once and future universe," *National Geographic* 163 (1983):704, and Rothman, "Cosmic lithium suggests the universe is open," *Scientific American* 261 (August 1989):16.
5. Woosley and Weaver, "The great supernova of 1987," *Scientific American* 261 (August 1989):36.
6. Weinberg, *The First Three Minutes*, p. 154.
7. Whitrow, *The Structure and Evolution of the Universe*, and Einstein, *The Principle of Relativity.*

CHAPTER 5

Needed: A Big Universe

■■■■■■■■■■■■■■■■■■■■■■

The Sun appeared on the fourth day. Rabbi Abahu said: "From this we learn that during the first three days, the Holy One Blessed Be His Name used to create and destroy worlds."

—Genesis Rabba

"In the beginning God created the heavens and the earth" (Genesis 1:1); from absolute nothing. "For six days God made the heavens and the earth" (Exodus 20:11); from that which was created in the beginning.

—Nahmanides,
Commentary on Genesis

The heavy elements did not originate in the Big Bang. They have been made over eons of time in the cores of stars from the hydrogen and helium formed in the Big Bang.

—Burbridge, Fowler, and Hoyle

Sages and scientists agree. The universe is getting bigger all the time. But until the 1920s, the prevailing concept was one of a static universe. Both philosophy and cosmology agreed. The belief was so strongly entrenched that Einstein put into his equations of relativity the fudge factor he referred to as the "cosmological constant" to force the prediction of a static universe. Theology, we have learned, saw the universe differently. Then and now, it views our cosmos as an expanding and evolving universe. Science has come to agree.

Now if, as theologians often claim, the universe is to provide mankind with an environment wherein he can work toward his and the world's improvement, why did God make a large universe? Why not just the Earth or our solar system?

God might have plunked man down in a world that was ready-made from the instant of creation. But that was not on the Creator's agenda. There was a sequence of events, a development in the world, which led to conditions suitable for man. This is evident from the literal text of Genesis 1:1–31. By God's time frame, the sequence took six days. By our frame, it took billions of years. Regardless, there was a series of events separated by time. At the end of that sequence, man was formed.

Now if there is anything in this universe for which we do not have an "inkling," it is the ultimate goal of the Creator. Erroneous notions regarding this goal often stem from the misconception that all existence exists for man alone. The foible in this perception of the universe is the failure to realize that existence itself is good. The Five Books of Moses are bracketed by explicit statements of the worth of being. At the start we are told: "And God saw all that was made and behold it was very good" (Gen. 1:31). Ultimately, in the poem that Moses tells to Israel on the last day of his life on Earth, we read, "The Rock [a synonym for God], His work is perfect" (Deut. 32:4).

Whatever God's goal may be for the universe, it does appear that there are certain ground rules that are used to reach it. One of those rules is that interactions in the physical world follow specific laws of physics and chemistry. When you put salt in pure, warm water, it dissolves. Always. When an energetic neutron strikes the nucleus of a uranium atom, the result is the fission of the nucleus. Always. If you want to develop life in our universe and do it according to the laws of nature as we know them, then you need a big universe. Always.

Always because the way the matter of our universe interacts is not haphazard. The recurring patterns of the chemical and physical reactions that we observe on a large scale are the results of forces operating at the level of the elementary particles. These forces appear as fixed and inherent properties of the substance from which all the universe is constructed. It also appears that these properties have been a part of the universe at least for billions of years and probably since its creation. We surmise this from the fact that the spectral properties of light received from galaxies billions of light-years distant are the same as the spectra of light originating in our solar system and galaxy.

Light is the result of electromagnetic energy emissions caused by changes in the energy levels of electrons. These energy levels are directly related to the specific chemical state of the atom. The similarity among spectra implies that the nuclear, atomic, and molecular structure of atoms in our part of the universe is extant throughout the universe. The light from some of the galaxies included in these spectral studies started its journey toward our current place in the universe billions of years ago, at times in so distant a past that the Earth was not yet even a molten agglomerate in a newly forming solar system.

Considering this, it appears that the forces that organize orderly electron positions about the nuclei of atoms have always functioned. The atomic structure appears to be an intimate and constant part of the matter of our universe.

We do not yet understand what regulates the laws that govern the orderliness of nuclear and atomic structure but we can describe their effects. Newton formulated the first of the fundamental statements. It is so common that we take it for granted. Matter is attracted to matter by something we have labeled gravity. The property of matter that generates gravity and allows it to propagate across the vast reaches of space is not known, but without it the nebulous gases of space would not have coalesced to form galaxies and stellar systems and mankind.

A nuclear force allows, or rather forces, neutrons and protons to bind and form atomic nuclei with the specific and regular structures of the 92 naturally occurring elements of our universe. A separate nuclear force causes certain types of nuclei to break apart spontaneously. We call this action *spontaneous fission*.

The electromagnetic force holds electrons in fixed and discrete atomic orbits. In doing so, this creates the orderly chemistry that allows individual atoms to group into molecules and molecules into coherent bodies in a predictable manner. The orderliness of these forces of nature made it *impossible*, in a natural way, to form galaxies and a solar system and mankind just after the Big Bang.

When the universe was very young, it was also very small. All the energy that today is spread over the reaches of space was concentrated into that confined, primordial volume. The substance of the early universe consisted mainly of high energy photons and neutrinos plus, relatively speaking, a tiny amount of matter in the form of individual protons, neutrons, and electrons. Photons and neutrinos have zero rest mass. Protons and neutrons have rest masses 1836 and 1838 times the rest mass of an electron, respectively. The electron is the lightest of the known particles. At rest, it weighs a mere 10^{-27} grams!

A single proton is what we refer to as a nucleus of hydrogen. Hydrogen is the lightest element. Heavier elements are formed by fusion of groups of the lighter elements. Helium, the second lightest element, is formed by the fusion of two protons and two neutrons. The mass or weight of the resultant helium is a bit less than the sum of the four particle masses from which it formed. This difference in mass appears as energy released as one product of the fusion reaction that changes hydrogen into helium.

Had the hydrogen been able to fuse into helium in the high-temperature (and, therefore, high-energy) conditions of the very early universe, our universe would be a different place than it is today. To reach helium, a series of nuclear reactions must occur in rapid succession. In each of these reactions, an intermediate product is formed. The first such intermediate is deuterium. It consists of a hydrogen nucleus with an added neutron.

While the helium nucleus has a form such that the neutrons and protons are tightly bound to one another, the bond of a single proton and a single neutron (that is, the deuterium nucleus) is not strong. The concentration of energy was so great in the early universe that whenever a nucleus of deuterium formed, it was immediately blasted apart by photons, the dominant entity of this early period.

As the universe expanded, temperatures and energies fell. It took almost four minutes after the beginning for the universe to cool to a temperature that allowed deuterium to withstand the pressure of photon bombardment. Hydrogen then rapidly fused into deuterium and then into helium. Helium (two protons and two neutrons) might then absorb more protons and neutrons, and in so doing build the heavier elements. There was, however, one fortunate circumstance: Nuclei with five or eight particles are highly unstable. To get beyond the nuclear weight of four mass units (the two protons plus two neutrons of helium), nucleides such as helium and deuterium must collide.

Time had taken its toll. The size of the universe was no longer measured in terms of mustard grains. The density of particles in the expanded universe was so low that random collisions of nuclei had become rare. The result: Little or no heavy element formation occurred at that time.

That might have been the end of the story had it not been for gravity. When comparing the strength of the inherent forces of matter at small distances, gravity is the weakest of the forces. But as the universe expanded, close encounters among particles became increasingly rare. The nuclear and electromagnetic forces are ineffective when separations exceed molecular dimensions. Gravity became dominant, affecting the relations among the groupings of mass in the universe.

Several hundred thousand years passed. Temperatures and photon energies had continued to fall in proportion with the universe's expansion. When the temperature fell below 3000°K, a critical event occurred: Light separated from matter and emerged from the darkness of the universe.

> And the Earth was without form and void and darkness was on the face of the deep, and God's wind hovered on the face of the waters. And God said: "Let there be light." and there was light. And God saw the light, that it was good; and God separated between the light and between the darkness
>
> (Gen. 1:2–4).

The photon energy at 3000°K is approximately 1 electron volt. This is approximately the minimum energy required to blast an electron in an atomic orbit out of that orbit. When photon energies fell below 1 electron volt, the photons were no longer sufficiently powerful to eject the atomic electrons. Electrons were at once drawn into stable orbits around the hydrogen and helium nuclei by the electromagnetic charges of these nuclei. Prior to that time, there had been a mix of photons and free electrons, or light and matter, in a turmoil of continual collisions. These photon-electron collisions had been so frequent that the photons (light itself) had been literally held within the mass of the universe. The situation changed dramatically as this soup of chaotic collisions was then suddenly cleared of the free electrons. With the electrons removed from this soup, by being bound in stable atomic orbits, the photons could travel freely. At that moment they were able to separate from the matter of the universe.

There are a whole range of photon energies, from the weak microwaves in a kitchen microwave oven to energetic gamma rays shooting from the Sun. But the freeing of photons from matter was destined to occur within that narrow range of energies we refer to as visible light. It was destined to occur there because our eyes see as light that most important class of interaction: the interaction between photons and orbital electrons. It is the smashing of photons into electrons in orbit around nuclei of atoms that gives rise to the color and visual texture of things. It is what lets us identify the ripe apple at the top of the tree.

The "light" of Genesis 1:3 existed prior to the Divine separation of light from darkness, which is described in Genesis 1:4. Both the Talmud and cosmology acknowledge that this first "light" was of a nature so powerful that it would not have been visible by humans. We have learned from science that the "light" of that early period was in the energy range of gamma rays, an energy far in excess of that which is visible to the eye. As the thermal energy of the photons fell to 3000°K, thus allowing electrons to bind in stable orbits around hydrogen and helium nuclei, not only did the photons break free from the matter of the universe ("separated" in the terms of the Torah), but they became visible as well. Light was now light and darkness dark, theologically and scientifically.

With an understanding that light was actually held within the primeval mass until being freed by the binding of electrons into atomic orbits, the enigmatic division by God between light (which is totally composed of photons) and darkness takes on a significance consistent with its literal meaning.

> . . . and God separated between the light and the darkness
>
> (Gen. 1:4).

*   *   *

With a premonition of the eventual findings of science, Nahmanides explained that from Genesis 1:4, and thereafter in the biblical text, the terms "light" and "darkness" refer to the phenomena as perceived by mankind.[1] That is, light is light and darkness is the absence of light. The darkness of verse 2 was not merely an absence of light. In these verses, it includes the meaning of the elemental source of energy. This very darkness, the sages say, contained the source of energy that was to power the forces that led to life. Furthermore, Isaiah 45:7 tells us that hoshek (Hebrew for "dark") is not merely the absence of light. It is a created, possibly the created, substance of the universe: "I [God] form light and create darkness [hoshek]." According to this verse, it is darkness, not light, that was created. The darkness was a black fire; a type of energy that emitted no light, just as the surface of the universe was black as long as photons and free electrons were mixed in a confused turmoil of energetic collisions.

When the matter of the universe was freed from the constant bombardment by photons, this matter, now consisting of approximately 75 percent hydrogen and 25 percent helium, could begin to cluster, to form galaxies and stars. But life as we know it could not yet form. Life requires more than hydrogen and helium. All those elements necessary for life, carbon, nitrogen, iron, iodine, and the really heavy elements such as gold and uranium, did not yet exist. It was a universe of hydrogen, helium, photons, and neutrinos. There had to be further synthesis of elements, but the expansion of space had reduced the particle density to such a low level that heavier nuclei could no longer form.

As eons passed, the needed conditions for element synthesis were established by the forces of gravity. The mutual gravitational attraction among the nebulous gases of primeval hydrogen and helium slowly formed clusters. As pressures within the cores of these clusters increased, the nuclei of hydrogen and helium were squeezed ever more closely together. Fusion among the nuclei started. The nuclear furnaces that still dot our night sky with light, the stars, had ignited. Within stellar cores the concentrations of hydrogen and helium, formed in the early universe, are so high and pressures so great that fusing of groups of light atoms is possible. From this stellar fusion came (and still comes) all the elements present in our universe other than hydrogen and helium. They are built just as a child joins together clusters of Lego blocks to form ever larger structures until finally the structure gets so large that it breaks.

Shortly following the Big Bang, the matter of the universe existed as a single massive nebula. The development from this nebula to the formation of stars and galaxies took time. How long it took for the first stars to form is anybody's guess. But we know it took not less than billions of years. And still the newly formed elements were locked within the stars. When the nuclear fuel of a star, the hydrogen and helium within it, is consumed, having been fused into heavier elements, the release of energy ceases. This energy had supported the outer layers of the star, and without this support, the star collapses. The plunge of the entire mass of the star toward its core releases a burst of energy that rebounds from the center as a massive shock wave. The outer layers of the star are shattered. They spew their supply of newly formed elements into space. Over eons, these elements form once again into new stellar systems, recycling the matter of space. Our solar system is one example of this recycling. We are made of recycled star dust. There is no other way—consistent with our understanding of cosmological processes—to account for the abundance of elements heavier than helium in our universe.

If we proceed according to the laws of physics and chemistry as they exist in our world, then preparing the elements of our solar system took many billions of years in the universe-based reference frame of time or into the third day by God's reference frame. All this time the universe was expanding. By the time the universe was ready for our solar system with its supply of life-essential light and heavy elements, it was already very big and very old.

As Rabbi Abahu, a fifth-century Hebrew sage taught, "From this [the fact that the Sun appeared on the fourth day] we learn that during the first three days the Holy One . . . used to create and destroy worlds."[2] Remember, these commentaries were not composed in response to cosmological discoveries as attempts to force an agreement between theology and cosmology. These commentaries, which are now paralleled so closely by the findings of astrophysics, stand as unchanging markers in biblical scholarship's view of our early universe. Before radioastronomy and spectrophotometry, could you expect Rabbi Abahu to talk of recycling of helium in stellar cores? Of course not. The language of science had not yet caught up with the language of man.

In the beginning God created *the* heavens and *the* earth"

(Gen. 1:1).

The redundancy in Genesis 1:1 of a simple two-letter Hebrew word *et*, meaning "the," leads the Talmud[3] to teach, "In the initial act of creation, the potential existed for all that ever will be contained in the universe." As a confirmation of this single act of creation, Isaiah 48:13 declares, "Also my hand laid the foundation of the earth and my right hand spanned the heavens; I call to them and they stand together." Time was needed to form that potential of creation into substantive matter. This is revealed by the slight, but important variation in the literal reading of two biblical verses. Compare carefully the wording of the texts:

In the beginning God *created* the heavens and the earth

(Gen 1:1).

For six days God *made* the heavens and the earth

(Exod. 20:11).

What do you find? All was *created* "in the beginning." But it took six days of *making* from the creation to form "the heavens and the earth, the seas and all that is within them" (Exod. 20:11). Six days in God's space-time reference frame and 15 billion years in ours.

Based on our understanding of the conditions just after the Big Bang, the laws of physics would predict either a universe composed predominantly of heavy elements, or a universe filled with matter so widely dispersed that stars and galaxies might never have formed. Forces are required that are not observed today. We find them only as characters written on a theoretical physicist's blackboard. To arrive at our universe from the conditions and substance that existed in the first millionth billionth billionth

billionth of a second after the Big Bang required a one-time ho-mogenization, which would direct the course of the universe's expansion. The first theoretical statement of this unique homog-enization was formulated in 1979 by Alan Guth at M.I.T. He referred to the concept as *inflation*.[4]

At 10^{-35} seconds after the beginning, the universe had a di-ameter of 10^{-24} centimeters. At that instant, a unique, *one-time* force—a sort of antigravity—developed. This force, acting for a minuscule fraction of a second, caused an expansion of the uni-verse at a rate far in excess of any rate prior to, or after, this episode. In this brief epoch, the universe inflated to the size of a grapefruit.

The biblical allusion to this one-time inflation is found in Gen-esis 1:2. "And darkness was on the face of the deep [the primeval space created at the beginning], and a wind of God [a one-time force mentioned only here in all of Genesis] moved on the face of the water [the common stuff from which the heavens, the earth, and all that they contain would be produced]."

The introduction into cosmological theory of this brief and rapid expansion of the universe is like a marriage of theology and science. In that first spot of time following creation, all the mass of the universal energy-matter continuum was concentrated in a single point. When a vast amount of mass is in a small volume, the gravity generated by this mass can be so great that nothing can escape. Not even light. It's what cosmologists call a *black hole*. Black holes were discovered two decades ago, notwith-standing the 3,400-year-old statement of Genesis 1:2. Matter, light, whatever comes near, is drawn into a black hole. They are formed today when a star with a mass about ten times greater than that of the Sun has used up most of its nuclear fuel. The cooling process that ensues allows the core mass to collapse toward the central point of the star. The concentration of matter becomes so great that its gravity pulls all the mass, photons included, into the center. Nothing escapes. It is black.

When, immediately following creation, all matter was concen-trated at one point, conditions existed for a super black hole. And indeed "darkness was on the face of the deep." Then according to Genesis what happened? Note the structure of the text carefully. The following is the literal English equivalent of the original

Hebrew text: "and darkness was on the face of the deep, and God's wind stirred on the face of the waters." There is no break for a new sentence although there is a grammatical pause, the equivalent of a comma in English, between the statements of darkness and the wind of God. A force, this wind of God, was required to start motion, the expansion of the black hole, which was the entire universe.

The "wind of God" or "God's wind" of verse 2 is not like the "strong east wind" that opened the Sea of Reeds for the Israelites leaving Egypt (Exod. 14:21). "And Moses stretched out his hand over the sea and God caused the sea to go back by a strong east wind all the night, and made the sea dry land, and the water was divided." At the exodus, the description is of a distinct physical, albeit God-inspired, phenomenon. It was a wind composed of moving air. Here in Genesis, Maimonides explains that had a true wind been meant, the action attributed to this wind (merahefet in Hebrew) would be illogical.[5] The Hebrew word merahefet means "to hover above," as a bird hovers above its nested young (see Deut. 32:11). It does not mean to blow. The wind of the "wind of God" has the meaning of spirit (Eccles. 12:7), or Divine inspiration (Num. 11:17), or God's will (Isa. 19:3). It is of great significance that this term wind of God is used only once in Genesis. It is a one-time phenomenon.

Astrophysicists also have no conventional explanation for what could have started the outward flow of matter. But very early in the life of our universe, they call for a one-time, new type of force, an "inflationary epoch."

This is what makes the comparison between theology and cosmology so interesting in this breaking of light out of darkness. One-time phenomena are almost never called on by physicists. It is too much like using a fudge factor. Yet both science and the Bible call on such a phenomenon at this juncture in our history. It was a wind of God, a ruach elokim in Hebrew, and an inflationary epoch in scientific terminology. It was needed and it occurred.

Both science and theology agree that the inflationary epoch did not bring light to the scene at once. Biblically we learn this from the fact that a new sentence is started to record the appearance of light (Gen. 1:3). This is the first biblical mention of light. Cosmology has taught the same fact in terms of photons being locked in a confused turmoil of collisions with the free electrons. Before light could break out from this massive nebula, scattering among free electrons and photons would have to be greatly reduced. That process took time.

With the separation of light from matter, matter could start to coalesce. Millions or billions of years passed. They fit within a day in the time frame of the Creator. Diffuse matter clustered to form galaxies and stars. Oases of light appeared in the vast expanse of dark space. Stars and planets and life were on their way.

> And God saw the light, that it was good; and God separated between the light and the darkness
>
> (Gen. 1:4).

Notes

1. Nahmanides, *Commentary on the Torah*, Genesis 1:4.
2. Genesis Rabbah, *A commentary on the book of Genesis*, redacted in the fifth century.
3. *Babylonian Talmud*, Section Haggigah 12a.
4. Guth and Steinhardt, "The inflationary universe," *Scientific American* 250 (May 1984):116.
5. Maimonides, *The Guide for the Perplexed*, part 1, chapter 40.

CHAPTER 6

Evening and Morning
Taking Order out of Chaos

■■■■■■■■■■■■■■■■■■■■■■

And God saw everything that he had made and behold it was a unified order.

—Genesis 1:31, Onkelos translation

A nd there was evening and there was morning." With only slight variations in the wording, the passage of each of the six days of Genesis is marked by this phrase.

Commentators on the opening verses of Genesis have wondered at the occurrence of the evening and morning sequence and the stated passage of each "day." This is especially true of the first few days of Genesis when, even in accord with a literal reading of the text, there was neither Sun nor Earth by which to reckon this flow of time.

The sequence in which a day starts with the evening follows the biblical concept of a day. Such is the explicit description of Yom Kippur, the Day of Atonement (Lev. 23:32). "It shall be for you as a sabbath of solemn rest . . . from evening to evening you shall celebrate your sabbath." The holiday begins with the evening and extends through the following evening. The puzzle remains as why to cite evening and morning, which are earthly concepts, for the first few days in Genesis. During those first days the literal text tells us that there was no Earth!

Among the ancient biblical scholars, there was a well-developed understanding of the phenomena that produced the day–night cycle on Earth. Nahmanides summarized this knowledge. "On the Earth both evening and morning are always present. There are on the Earth at every moment ever changing places where it is morning and in the places opposite them it is evening."[1] This reveals quite an exact comprehension of the illumination of the Sun's light on a spherical Earth for a time in the history of mankind when most of Western humanity dreaded falling off the edge of a flat Earth.

Nahmanides declares that each of the six days of Genesis had 24 hours' duration. As we learned in chapter 2, this refers to a space-time reference frame of the Creator. Even if each period was for a day (albeit a godly day), why the daily repetition of the phrase, "and there was evening and there was morning"?

Nahmanides, backed by Onkelos, sweeps away the conundrum with a brilliant insight into the language of the Bible. His interpretation flows from a transcendent knowledge of Hebrew. As logic would dictate, Nahmanides sought the answer to these questions in the biblical text itself. He found the answers in the order of the words and in their Hebrew roots. With each interpretation we return to the underlying principle: The Bible talks in the language of man, the average man. The sages had to dig to get the deeper meanings.

... and there was evening and there was morning ...

The Hebrew word for "evening" is erev. This is the literal meaning of the word, although the root of erev carries with it implications far beyond that of a setting Sun. What is the visual sensation of evening? Darkness begins. Objects become obscure, blurred. The root of erev means just that, "mixed-up, stirred together, disorderly."

The Hebrew for "morning" is boker. Its meaning is quite the opposite of erev. Morning brings the first light. Objects, visually mingled by the dark of night, become distinct entities and this is the root meaning of boker, "discernible, able to be distinguished, orderly."

Had the text said "and there was morning and there was evening," our concept of a day might have been better satisfied. The sequence would have at least included the light of the day. But had the text followed this human logic, it would have forfeited its cosmic message. The text is telling us something crucial about the flow of matter in this universe, something that can only happen to a subsystem contained within, and in contact with, a larger system. The phenomenon was so important that it was identified six times as the flow from evening to morning. We are being told that within this parcel of space where mankind was to stake his first roots, there was a systematic flow from disorder—chaos or "evening"—to order—cosmos or "morning."

To appreciate the rarity of this flow from disorder to order, we must at this point acquire a fuller understanding of the laws of thermodynamics and specifically of entropy.

ENTROPY: CHAOS OUT OF ORDER

The order that we see about us, the exquisite orchestration of inorganic chemicals and organic molecules into a symphony of life is misleading if we apply it to the universe as a whole. Those intricate and regular arrangements of matter that are characteristic of life stand in sharp contrast to the ever-increasing chaos of the universe, a chaos that at times is referred to as the impending "heat death" of the universe. For with each act, be it as simple as a sustained nuclear reaction, or as complex as a parent tickling a giggling child, some energy or some matter or both are transferred from an orderly state of potential usefulness to a random, chaotic state from which useful work can never again be drawn.

The energy level of a system, be it a bacterium or a galaxy, is describable by statements of its thermodynamic properties. These statements, grouped into what are referred to as the three *Laws of Thermodynamics*, have been found to be valid for all conditions yet observed.

The first law states that regardless of the reactions taking place within a closed system, the sum of the energy and mass of the system remains constant. This law is observed every day of our lives. When hot water is mixed with cold water in an insulated container, the hot gets cooler and the cold gets warmer but the total heat remains the same. The molecules of the hot water were moving rapidly and randomly within the liquid. The molecules of the cold water were also moving randomly but more slowly. The speed of this molecular motion, the kinetic energy of the molecules, is what determines the temperature of an object. Molecular motion is in fact what heat is. When these fast hot molecules were mixed with the slower moving molecules of the cold water, random collisions among the two groups soon distributed the energy equally among them so that all were eventually in moderate motion. This moderate motion gives us, on touching the water, the sensation of coolness. In the mixed water, the initial high energy of the hot water has been shared and literally diluted by the low energy of the cold water, but the total amount of heat or energy of the system has not changed.

The first law of thermodynamics provides the groundwork for what is most relevant to our discussion here: the law of increasing entropy. This is embodied in the second law of thermodynamics. A simple way of remembering the laws of thermodynamics is as follows: If the first law states, You can't win, then the second law states, You can't break even. (The third law is You can't get out of the game!)

Having mixed the cup of hot water with the cup of cold water, there is no way we can separate the now lukewarm molecules back into two separate groups, one very hot and one very cold, without the expenditure of more energy. We can recover that useful form of energy represented by the very hot water only by adding more energy to the system. With every action, chemical or physical, and nuclear as well, there is a decrease in the amount of energy useful for future actions.

In the Sun, nuclear reactions convert solar mass into energy, which travels through space to warm the Earth. There is no efficient way of converting the lower warmth of the Earth back into the elements that produced the initial nuclear reactions. Within the myriad on myriad of stars of the universe, mass is being propelled outward as radiant energy. The sum of the mass plus energy of our universe remains unchanged; that is the first law of thermodynamics. But the cosmic potential to produce useful energy in the form of high-temperature heat is ever diminishing. Stars that once burned at temperatures far in excess of those of the Sun are now frigid lumps of iron, their nuclear energy expended.

Based on our understanding of physics, the universe is tending toward a frigid temperature of $-273°C$. When this temperature has been reached, no further work will be possible, because there will be no temperature differences to induce a flow of heat. It is the flow of heat, from hot to cold, that provides the basis for work. Without the potential for work, the heat death of our universe will have occurred.

The flow from concentrated, organized forms of energy toward more randomly, disorganized energy distributions is referred to as an increase in the entropy of the system. We see randomness or entropy increase in every observable system. The trend in the universe is toward chaos, not toward cosmos.

And yet within a limited portion of the universe, highly ordered structures have appeared. We humans are the most extreme example of such localized order. Our life systems, or even those of the most simple, single-celled bacterium or alga, are so vastly complex that one might question the validity of the second law of thermodynamics. Only when we view life processes in context with their environment does the overall increase in entropy become apparent.

The Sun showers a pond with solar energy. To form the orderly proteins and carbohydrates of an algal cell from the confusion of inorganic molecules in the surrounding water, the alga absorbs the photon energy of light and converts it into chemical energy concentrated in the molecule ferredoxin. The ferredoxin, in turn, contributes its energy to reduce carbon of carbon dioxide and combine it with water to form the sugar glucose. Each reaction produces heat as a waste product.

We are no different. We use the heat that is a by-product of the biochemical reactions of our bodies to keep us at the temperature that is optimum for the biochemistry of our bodies: 98°F. But stand still in a room with a temperature of 98°F and the sweat rolls off. Why? If our bodies struggle in cold weather to keep our inner organs at 98°F, shouldn't we be at ease in 98°F air? Except for the inevitableness of entropy's increase, we might be. Our metabolic processes, even when we are at rest, produce so much extra heat, never to be recovered, that we have developed an entire system, that of perspiration, to remove this excess heat by evaporative cooling.

Now comes the major physiophilosophical question: If the universal trend is toward chaos, why is there order at all? Why doesn't everything tend toward disorder?

The ordered complexity of life is clearly extraordinary, but even inorganic matter has impressive order. The salt that forms from the waters of the Dead Sea, when viewed with sufficient magnification, has a regular and orderly crystalline structure. So do quartz and snow.

At the atomic level there is even greater regularity. All atoms with the property of carbon have 6 protons in their nucleus; that is what defines them as carbon. All uranium atoms have 92 protons. When we burn sodium, the light emitted always has specific frequencies of colors. A carbon flame has different, but equally fixed, spectral lines of emission. These spectral radiations are the result of the fixed and predictable orbits that the electrons have in each type of atom.

The spectral pattern of a given element is the same here on Earth as in the light of galaxies so distant that billions of years passed between its emission from these galaxies and its reception in the eyes and telescopes of mankind. This spatial and temporal consistency in spectral patterns implies a consistency in the laws of physics on a universal scale.

The inherent properties of matter that permit, or rather cause, the orderly organization of the matter of our universe are represented by the strong and weak nuclear forces, the electromagnetic force and the gravitational force. The universe might not have included these forces. If such were the case, then chaos rather than cosmos would indeed have ruled for matter just as it does for heat. For while matter can be self-organizing, heat lacks this property.

The universe might have remained the mass of randomly moving electrons, protons, and assorted subatomic particles present at that first speck of time we refer to as "the beginning." If such were the case, we would not be here to wonder whether there is an overriding purpose to the flow of the universe from its state at the Big Bang to the form we observe today. The author of Genesis thought this flow toward order was sufficiently important and exceptional to emphasize it by the regular repetition: "and there was evening and there was morning."

The compartmentalization of the events of our genesis into days bracketed by erev and boker, or evening and morning, is convenient for talking in the language of man. But the root meanings of the words hold the secret. Let's put what we have discussed about the expanding universe and entropy together with the root meanings of *erev* and *boker*. It makes a very sensible package.

In the beginning there was turmoil, a cosmic soup of highly concentrated energy and matter all mixed together. There were neither heavy elements ready to release the heat of radioactive decay nor energetic chemical complexes available to power the processes of life. Although matter was in a state of tumultuous chaos, the energy of the universe was in a condition of low entropy because it was so highly concentrated. This high concentration was represented by the $10^{32}°K$ temperature. The potential to extract work from this heat has never again been as great and, hence, the entropy has never been as low as at that instant. That initial concentration of energy has powered the universe since its creation.

With the expansion of the universe, the energy concentration fell. Temperatures dropped drastically. As we have learned, the average photon temperature today is a mere $3.5°K$. Entropy increased throughout the volume of the universe, but this increase was not uniformly distributed. At many places within the increasing volume of the universe, local order was imposed on matter by those basic properties instilled within all matter at the creation: the nuclear, electromagnetic, and gravitational forces. Matter started to join with matter. Light and matter became distinct.

When the supercompression of mass and energy exploded into the Big Bang, the energy contained within the universal core was converted partly into kinetic energy represented by the flight of mass out from the central core and partly into increasing potential energy represented by the separation among the masses as these masses became dispersed through space. Gravity has let us cash in the potential energy. Under its influence, a diffuse nebula of galactic gases can draw together and coalesce to form distinct galaxies and stars.

The converging flow of matter toward local centers of concentration produces high velocities and eventually high pressures. As the matter crashes toward a local center, both the velocity and pressure become expressed as heat. Temperatures in the core rise. At approximately a million degrees Kelvin, the velocity of nuclei becomes sufficiently high to cause nuclear fusion on collision with other nuclei. Hydrogen converts to helium and helium to heavier elements. In the process of fusion, some of the mass converts to pure energy. A nuclear furnace has ignited. A star has been born. At such localized points in space, very high temperatures have been regained in the otherwise now-cold universe.

In a sense, the Big Bang put its energy into spreading matter. Gravity helped recover part of that energy by forming second- and third-generation hot spots, which we call stars. Life's processes derive their power from this recycling of energy and matter.

The biblical text describes this localized progression from less order to more order as a flow from evening to morning, or more accurately from *erev* to *boker*. Hundreds of years before the appearance of the Greek language, before the words *chaos* and *cosmos* were first written, the step-by-step progress toward an orderly world was described by the subtle language of evening and morning.

There is an ancient text that confirms this interpretation. It is the Onkelos translation of Genesis.

The recurring term "it was good" is used in the biblical description of each of the six days of Genesis when a process of formation of a particular entity or entities was completed and had resulted in something durable. Only during the second day is the term "it was good" missing. On this day the firmament produced a division of the waters, but it was not until the third day that the lower waters were gathered into seas. At this stage "it was good" (Gen. 1:10).

In the first 30 verses of the Bible, Onkelos translated each phrase "and it was good" literally. On the sixth day, at the end of the description of the making of the universe, the literal translation of the Hebrew text states: "and it was very good" (Gen. 1:31). Except for the addition of *very*, the text is identical to the previous phrasing. Onkelos, however, based on an unknown source, made an exceptional, radical, and quite extraordinary departure from the trend he had set. In Genesis 1:31, Onkelos interpreted "and it was very good" as "and it was a unified order."[2] Writing 1800 years ago, Onkelos realized that the progression described in the opening chapter of Genesis as a sequential transition from evening to morning was much more than a quantitative advance that might have been measured as increases in the number of stars or the species of animals present in the universe as it progressed from okay to good to very good. Rather Onkelos perceived that a *qualitative* change in the nature of the universe had occurred, a change from disorder without the possibility of life to an order with life.

> And God said, let us make mankind in our image, after our likeness; and let them have dominion over the fish of the sea and over the birds of the heavens, and over the cattle and over all the earth, and over all the creatures that move upon the earth. And God created mankind in his image, in the image of God created him, male and female created them. And God blessed them and said to them be fruitful and multiply and fill the earth and subdue it and rule over the fish of the sea and the fowl of the heavens and over all living creatures that move upon the earth.... And God saw all that he had made and behold it was very good and there was evening and there was morning, the sixth day
>
> (Gen. 1:26–28, 31).

Notes

1. Nahmanides, *Commentary on the Torah*, Genesis 1:5.
2. Onkelos, *Translation of the Torah*, Genesis 1:31.

CHAPTER 7

Odds
The Chance of a Lifetime

■■■■■■■■■■■■■■■■■■■■■■

Regarding the origin of life: Time performs the miracles. Time is in fact the hero of the plot.

—Nobel laureate George Wald

Random events cannot account for the origin of life, at least not in the time available.

—Harold Morowitz

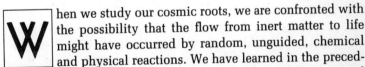

hen we study our cosmic roots, we are confronted with the possibility that the flow from inert matter to life might have occurred by random, unguided, chemical and physical reactions. We have learned in the preceding chapters that physical and chemical reactions follow natural laws, which were established with the creation of the universe. If we believe that this creation was the result of an act by God, then we might acknowledge a random development of life as being God-given through the Divine origin of the laws of nature. After all, it would have been these God-given laws of nature that had governed the interactions that led to life. God and not chance would still be the force of our existence.

With or without God, scientists and philosophers have espoused randomness in the flight of time toward the target of life. They have claimed that given enough time, a multitude of random combinations of matter will inevitably result in life.

In fact, at times it seems that the living and the inert are not so very different. Growth, for example, is certainly not the private bailiwick of life. A massive lump of salt in the Dead Sea starts as a single crystal that gradually adds more salt to its surface, removing it from the environment and organizing it to form the regular shape of the salt crystal. Electric charges on the crystal surface induce this reaction by attracting the appropriate atom to each given location. Granted, the salt crystal grows only in the sense that it reproduces an attached structure (the crystal) nearly identical with the initial spatial arrangement of atoms, but the simple form of life known as the virus does something quite similar. It takes from the host cell, which it has invaded, the cellular material needed to make replicas of itself.

Amino acids serve as the building blocks of protein. In living organisms, these organic molecules are synthesized within the cells of plants and microbes. In 1953, Stanley Miller reported the results of an experiment in which he synthesized amino acids by totally nonbiological reactions among simple inorganic compounds found throughout the universe.[1] Miller showed that amino acids, the very basis of all life on Earth, could indeed be the product of random reactions. Their production did not depend on some supernatural guidance.

The machinery for organic cell reproduction is found in the genetic material referred to as DNA and RNA. We haven't a clue as to its origins, but it appears that it works in a manner consistent with laws of physical chemistry. The DNA and RNA select, from the pool of organic and inorganic compounds within a cell, those molecules that fit into the genetic-code material. While the DNA and RNA deal with the organization of complex molecules such as amino acids, qualitatively these acts of "reproduction" are as unintellectual as the growth of an inorganic mineral.[2]

The individual chemical and physical reactions involved in the maintenance of life are quite similar to those found in reactions among nonliving substances. Clearly the processes of the living and the inert overlap. The distinction between the two is one of quality, not of type.

Because life was brought forth from the Earth and the waters (Gen. 1:11, 20, 24), it is not surprising that the reactions of biological molecules maintain a resemblance to those of the inorganic precursors. Both are based on the inherent properties that govern all matter. Perhaps it is the organization of the myriads of these individual, "natural" reactions into a coherent rhythm that brings us to the concept of life. If that is the case, then the biochemical organizer, the genetic-code material, is a key to our differentiation between the living and the inert, or technically stated, the biotic and the abiotic.

The amino acids found in Miller's experiments were far from being alive. But even if his research had actually produced life, the question would still have remained: Can life arise from random reactions among inorganic molecules? Miller's experiments represent conditions that are not at all random. Each was life outside a test tube—the researcher—carefully manipulating the environment within the test tube.

Those who believe that life can arise from the nonliving are found on both sides of the theological aisle. One might argue that biblically, the forces of life were placed in matter at the creation. Thus on the third day (Gen. 1:11–12), the Earth was able to "bring forth" plant life. "And God said, 'Let the earth bring forth grass, herb yielding seed and fruit tree making fruit after its own type.' " Although the timing of the Earth's flowering into botanic abundance was in accord with God's command, there is no mention of creation on this day. The same is the case for the start of animal life. On the fourth day the heavens cleared, and on the fifth day the waters were filled with animal life. "And God said, 'Let the waters swarm with moving creatures having a living soul' " (Gen. 1:20). Again, no mention of creation. It might appear that the waters, as the Earth, had the inherent potential to produce life. That the Earth could "bring forth" both plant and animal life is clear from the same verb being used for the two acts:

And the earth brought forth grass . . .

(Gen. 1:12).

And God said, "Let the earth bring forth living creatures [literally "bring forth living souls"] after its own kind . . ."

(Gen. 1:24).

In 1860, the question of whether life could indeed arise spontaneously from inert matter was almost laid to rest by Louis Pasteur. For centuries prior to Pasteur's work, the phenomenon of spontaneous generation had been considered as a fact. Any food left for a few days swarmed with life. Christian priests, such as

Augustine, and scientists such as Newton, Harvey, and Descartes accepted its reality. But scientific inquiry had begun to produce results that raised serious questions as to its validity. In 1668, Francesco Redi, an Italian physician, had shown that maggots grew from the eggs of flies and not from spontaneous generation arising from rotting meat. A century later, with the development of the microscope, adherents to spontaneous generation received a new source of proof. Microbes, those tiny specks of life, were seen to come into existence when any source of nutrients was kept in the open for a few days. John Needham, a Scottish clergyman, and Lazzaro Spallanzani, an Italian abbot, argued over the efficacy of sterilization as a way of refuting spontaneous generation. Broth boiled and then sealed remained free of life. Broth boiled and left open soon rotted. Needham claimed that overextensive boiling had killed the life force, that vital force present in the molecules of all inorganic matter. The broth in the open flasks rotted because it had access to fresh air.

Pasteur sought a way to allow fresh air into the flasks without its potential charge of microbial life. He had observed that dust collected by drawing air through a wad of cotton literally teemed with life. To meet the objections of those committed to the idea of a "vital force" contained within inorganic matter, Pasteur either covered his flasks with wads of cotton or used flasks with long, narrow S-shaped necks. Pasteur proposed that although air could pass through the cotton and the narrow tubes, dust and microbes would impinge on and adhere to the solid surfaces of the cotton and the tube walls prior to reaching the nutrient broth. As the boiled broth cooled, the steam that filled the free volume of the flask condensed. Cool air was drawn into the flask. The filtering succeeded. The broth remained pure even in the presence of the "fresh" room air.

Pasteur had solved the riddle of the spontaneous generation of life. It was all from microbes in the air and fly eggs and not spontaneous at all.

Or was it?

The spontaneous generation of maggots from meat may have ascribed to meat a misplaced power, but that left the main question of life's origin without scientific proof. If spontaneous generation was invalid, what was the source of life? Did Pasteur's results include, as an implicit spin-off, a proof of the supernatural creation of life? An agnostic might reply to this, God forbid!

In 1936, Alexander Ivanovich Oparin, a Russian biochemist, published a book titled *The Origin of Life*.[3] In it, he described the conditions likely to have existed on the primitive Earth and the random chemical and physical processes possible in such an environment. These processes, he asserted, *inevitably* led to life. Seventeen years later, Stanley Miller used almost these same conditions in his experiment to produce amino acids. Oparin speculated and Miller proved that lightning and other sources of energy naturally present on Earth could convert inorganic molecules into several of the compounds present in life.

But how was nature to get these individual molecules organized into the complex array found even in the simplest forms of life? In theory, the needed sequence that would carry the basic molecules through the complex path ending in a true protein could occur step-by-step in chance reactions over long periods of time. The difficulty with such a slow and random process is that just as there is a given probability of forming an intermediate product in this chain of products leading to life, there is also a probability of its spontaneous dissolution.

At each step as we go from simple to more complex compounds, we are in a sense swimming upstream in the flow of entropy. The result is that the likelihood of the disintegration of a newly formed organic compound is much greater than the likelihood of its formation.

If destruction predominates over formation, how is it that living organisms regularly produce complex compounds and do so in copious amounts? Life does it by working in the highly protected environment within its cells, by using catalysts that have the ability to select and concentrate the needed chemicals and to increase rates and extents of reactions, and by expending considerable energy to accomplish the tasks. The protected environment needed by life is found within life itself.

From the simplest to the highest forms of life, if the cellular system fails, the organism dies. Its subsequent rapid decay is clear evidence for the chemical instability of the compounds from which life is composed.

The catalysts of living organisms, called enzymes, are themselves proteins produced by already-living cells. A reaction that may take seconds within an enzyme-driven, temperature-controlled 98°F system of an animal might take years or longer in an uncatalyzed system. Neither enzyme nor protective cell wall were available to the molecules that preceded life.

As we experience it, life is required to produce life.

The world-famous biologist George Wald saw no problem with this. Writing in *Scientific American*, Wald presented a convincing logical argument that indicated that random processes following the physical laws of our universe can and indeed did account for the spontaneous generation of life from the nonliving. For Wald, life is an *inevitable* product of chemistry. One only has to wait for the random events to occur. "Time itself performs the miracles," he wrote. "Time is in fact the hero of the plot."[4]

I dwell on the account of Wald's assumptions because there is an important lesson to be learned from his definitive statements. But the lesson is *not* that life is inevitable. The lesson is that the trend of thought in this controversial study of life's origin has very often been based on *poorly researched science presented as fact* by one or a few noted personalities. To understand the potential impact of Wald's opinions on the lay public, we must know his position in the community of scientists. By 1948 Wald was a full professor at Harvard. He is one of the pioneering researchers in the biochemistry of vision. For his research, he received the Nobel Prize.

Now what do you suppose the audience of *Scientific American* thinks when a scientist with these impeccable credentials publishes therein his "facts" on life's origins? That he knows what he is writing about? That it is a certainty that random events can and did lead to life? That it is a proven truth that evolutionary forces, guided solely by the laws of nature, led from the first microbe to man? These are certainly the impressions.

Unfortunately, Wald's skills in mathematics seem to be less than his skills in biology and it is on the mathematics of probability that Wald's rather unoriginal argument rests.

In 1968, Professor Harold Morowitz, a physicist at Yale University, published the book *Energy Flow in Biology*.[5] Along with other physicists and mathematicians, he had become concerned about the casualness with which some scientists studying the origins of life were assuming that unlikely events must have occurred. These scientists were making assumptions without any attempt to rigorously investigate the probability of such events. Morowitz presented computations of the time required for random chemical reactions to form a bacterium—not an organism as complex as a human, not even a flower, just a simple, single-celled bacterium. Basing his calculations on optimistically rapid rates of reactions, the calculated time for the bacterium to form exceeds not only the 4.5-billion-year age of the Earth, but also the entire 15-billion-year age of the universe. The likelihood of random processes producing life from a primordial bath of chemicals is even less likely than that of your shaking an omelet and having the yolk and the white separate back into the original form of the egg!

But the impression that chance was the source of life, planted by distinguished personalities such as Wald, remains with the general public even though *Scientific American* later acknowledged that Wald had erred.

Ironically, it is the fossil record itself that is gradually dispelling this argument of chance. You see, neither 15 billion years nor 4 billion years are available for this random development of life. Life, we are learning, appeared on the Earth almost immediately after the Earth formed!

If we are seeking the development of life on Earth, then some basic conditions must exist. The simplest is that the molten agglomerate that was the primordial Earth must have cooled to a temperature that allowed liquid water and hence aqueous chemical reactions to occur. This cooling of the Earth's crust is estimated to have occurred some 4.5 billion years ago. From that time onward, water may have been available. The appearance of sedimentary rocks provides evidence of liquid water because sedimentary rocks are formed from the weathering and erosion, usually by water, of other rocks. Sedimentary rocks first appeared 3.8 billion years ago. This may be the earliest time that aquatic chemical reactions could have occurred on a widespread scale.

If we are searching for fossil evidence of the earliest life, then we need conditions in which fossils can form. An organism lying among large rocks will not develop into a fossil. Fossils can be preserved only in rock that has weathered, sedimented, and then

resolidified with the fossil-to-be contained in the sediment prior
to its solidification.

As such, the first sedimentary rocks provide evidence both for
the first presence of abundant water and also for the earliest potential preservation of fossils.

The oldest sedimentary rocks found to date are in the southern
African shield and in the Canadian shield in North America. In
the midst of the African shield, there is chert rock known as the
Fig Tree Formation. Within these sediments, dated at over 3.3
billion years old, the fossils of spherical and rod-shaped single-
celled organisms were discovered by Elso Barghoorn and J. W.
Schopf of Harvard.[6] The dimensions of these fossils match those
of similarly shaped microbes found in present-day aquatic environments.

This earliest evidence of life is dated less than 500 million
years after the appearance of the first sedimentary rocks, the oldest
rock types able to contain fossils. At this relatively young age,
several forms of life had already formed. It is statistically improbable, in fact, essentially impossible, that random events produced this life in such relatively short time.

If random events leading to life are statistically not probable,
where does that leave the argument? For this I know of no definitive answer. Research has demonstrated that basic molecules
found in volcanic gases—methane, ammonia, and water vapor—
can form into those molecules we associate with proteins, the
amino acids. Furthermore, the biochemical reactions of the living
cell are, when viewed individually, quite similar to selected inorganic reactions. When viewed as a series of isolated chemical
reactions, life might be confused as an extension of the inanimate
world.

The stumbling block in this transition from the geochemical to
the biological is the appearance of the genetic code. The genetic
material of a cell does more than produce progeny. It provides
the matrix on which all the biochemical needs of a cell are struc-
tured, from extracting chemical energy held in ingested foods
through production of the hundreds of proteins needed for cellular maintenance, growth, and production, and it does this in
accord with the extraordinarily complex cadence of life.

The long helixlike molecules, deoxyribonucleic acid (known as DNA) and ribonucleic acid (RNA), are the basis of the genetic material of all living cells yet analyzed. It is perhaps because of the extraordinary complexity of the genetic material that all forms of life have the same type of genetic code. Whether we are studying a blade of grass, a bacterium, or a human, all base their existence on the same 20 amino acids. All the proteins formed by the genetic material of life are formed with the same left-hand configuration to the molecule. All the sugars have a right-hand molecular configuration. In inanimate matter, the configurations are random. Furthermore, in spite of the complexity of individual proteins, often containing specific combinations of as many as 300 individual amino acids, there are several nearly identical proteins that appear in most forms of life.

This equivalence among all life forms is strong evidence for a single source of all life. It is not plausible that this similarity arose by chance. There are 20 different types of amino acids used in forming proteins. The probability of duplicating, by chance, two identical protein chains, each with 100 amino acids, is 1 chance in 20^{100}, which equals the digit 1 followed by 130 zeros or 10^{130}. To give perspective to the extraordinary magnitude of this number, realize that there have been less than 10^{18} seconds in the 15 billion years since the Big Bang.

To reach the probable condition that a single protein might have developed by chance, we would need 10^{110} trials to have been completed each second since the start of time! To carry out these concurrent trials, the feed stock of the reactions would require 10^{90} grams of carbon. But the entire mass of the Earth (all elements combined) is only 6×10^{27} grams! In fact the 10^{90} exceeds by many billion times the estimated mass of the entire universe! With these odds, it becomes clear that chance cannot have been the driving force that produced similar proteins in bacteria and in humans. But there *are* similar proteins in bacteria and humans.

There was, it appears, only one ancestral model for all life. But perhaps many types of genetic code are viable, and the code of today just happens to be the one that formed first. If this is the case, then our code is just one of many possibilities and its random development is less improbable. But if many code types are viable, then more than one should have formed and survived. This is certainly true if we base our opinion on present life. Living organisms are found in every environment where nutrients are present, be it a gas vent 5 kilometers below the ocean surface or the arid climate of a desert. Diversity of life form is also common at all levels of complexity, from bacteria to mammals. It is nature's way to make use of all the available resources and to ensure the survival of life when and if calamity strikes. If diversity is a characteristic of all life, then this implies an expectation of diversity in the development of the genetic code as well. But this is not the case. Only one code exists very likely because only one code is viable.

As we know, fossil evidence for a variety of microbial life forms has been found in sedimentary rocks more than 3.3 billion years old. The oldest sedimentary rocks are dated at about 3.8 billion years. In this span of approximately 0.5 billion years, the common ancestor of life must have developed the extraordinary and exquisite chemistry of life and also mutated sufficiently to have produced a variety of progeny.

The appearance of life on Earth almost as soon as the Earth was able to host life and the improbability of a random development of our genetic code in the available time has led scientists in many disciplines to suggest extraterrestrial sources for life on Earth. Nobel laureates Svante August Arrhenius and Francis Crick and astronomer Sir Fred Hoyle are among those who have looked to space for our origins.[7]

Even if we find that extraterrestrial seeding was the source of life on Earth, this would not solve the question of the ultimate origin of life. As Morowitz pointed out, even 15 billion years are insufficient for unguided, random reactions to produce life.

What we observe in this search for our cosmic roots is that many respected scientists, working in a range of disciplines, are seeking forces, other than those usually observed on the Earth, to explain our fossil record because *the fossil record itself cannot be explained by the conventional laws of chemistry and biology.* There is a new awareness in the scientific community that the simple evolutionary approach of inorganic chemistry leading to the biochemical requires modification.

*   *   *

... and the Lord God formed man of the dust from the ground ...

(Gen. 2:7).

Both biblically and scientifically speaking, we are formed of
the stuff of the universe. A major step forward in our scientific
perception of the universe is that we now know that *random
events did not do the forming.*

Notes

1. For a discussion of the implications of the Miller paper see: Dickerson, "Chemical
   evolution and the origin of life," in Dobzhansky et al., *Evolution*, p. 30.
2. Taylor, *The Great Evolution Mystery*, pp. 166–169.
3. Oparin, *The Origin of Life*.
4. Wald, "The origin of life," *Scientific American* 191 (August 1954):48.
5. Morowitz, *Energy Flow in Biology*.
6. Barghoorn, "The oldest fossils," *Scientific American* 224 (May 1971):30; Schopf,
   "The evolution of the earliest cells," *Scientific American* 239 (September 1978):84.
7. Crick and Orgel, "Directed panspermia," *Icarus* 19 (1973):341; and Hoyle and
   Wickramasinghe, *The Origin of Life*.

CHAPTER 8

The Earth
A Cradle for Life

■■■■■■■■■■■■■■■■■■■■■■

To everything there is a season and a time for every purpose under heaven.

—Ecclesiastes 3:1

T he formation of stars and planets seems to be as inherent to the universe as the force of gravity. The biochemistry of life when considered as a series of individual reactions appears to be as natural as the laws of chemistry and physics. But the transition from inorganic matter to living organisms is so complex that, as we have learned, many respected and accomplished scientists have suggested a necessity for extraterrestrial seeding of life itself.

Whether the formation of life on Earth took place in isolation or with input from extraterrestrial sources is still unanswered. At the very least, it can be stated that the continuation of life on Earth and its successful development from the simple forms we find in 3.3 billion-year-old fossils to the complex organisms of today reflect the Earth's extraordinarily suitable conditions for life. It is as if the Earth and its position within the solar system were designed to be both womb and cradle for life as we know it.

The appropriateness of Earth as the setting in which life was to be developed is rooted in the formation of the solar system some 4.5 to 5 billion years ago. The mechanics of the formation are speculative. The saga may have occurred as follows.

In the spiral galaxy we call the Milky Way, two-thirds of the way out from the center (that is, about 30,000 light-years), a massive cloud of gas and dust-size particles of minerals and rock started to contract. The cloud contained the recycled debris of bygone supernovas. Now, with the products of stellar-core nuclear fusion, the cloud contained traces of all the elements of our universe. As the cloud collapsed toward its center, its rate of spin increased, much as the spin of a figure skater increases as she pulls her arms in toward her body. The centrifugal force of the spin flattened the cloud into a disk. Counter to this outward thrust, the attraction of gravity drew most of the matter to the central region where the Sun was to form.

Within the flattening disk, particles of rock and ice not drawn into the center were able to cluster and agglomerate. As these clusters grew to be planets, ices and frozen gases that had coated the surfaces of the solid particles were incorporated into the body of the planet.

Possibly because there was a temperature gradient within the fledgling solar system, those planets that formed closer to the Sun (Mercury, Venus, Earth, and Mars) were enriched in the less volatile matter such as rocky minerals. Planets farther out were enriched in the volatile gases, especially hydrogen and helium. Jupiter, the giant of planets, has such a massive amount of hydrogen that its atmospheric surface pressure is some 3000 times that of the Earth.

Some statistics provide a perspective for the physical structure of the solar system and the Earth's position in it (see Table 3). Considered separately, few of the values in Table 3 would indicate that the Earth is anything special. Taken together, the values are almost too good to be true if accommodation for life is the goal. It is as if the Earth were made to order for life's eventual appearance.

The almost-too-good-to-be-true nature of the Earth started within a second of the creation and it still continues today. In the very early universe, the homogenizing force (discussed in chapter 5) referred to biblically as "a wind of God" stirring on the face of the waters (Gen. 1:2), fine-tuned the distribution of energy and matter within the young, expanding universe. This allowed the matter of the universe to expand past a density hurdle (recall that in the beginning all the matter of the universe was compressed into a volume smaller than a grain of mustard) that might have fused the elementary matter into massive particles. The unenviable result would have been almost immediate universal collapse.

With expansion healthily on its way, light and matter were still

TABLE 3. THE SOLAR SYSTEM

	Distance from Sun (million km)	Mass (the Earth's value = 1)	Gravity at surface (the Earth's value = 1)	Atmospheric pressure at surface (the Earth's value = 1)	Density (water = 1)	Atmospheric composition	Surface temperature (°C)	Volcanoes
Sun	—	340,000.	30.	—	1.4	H, He	6,000	—
Mercury	60	0.1	0.4	—	5.6	none	—	unknown
Venus	110	0.8	0.9	90.	5.1	>95% CO_2	500	yes
Earth	150	1.0 (6×10^{24} kg)	1.0	1.0	5.5	80% N 20% O	15	yes
Mars	230	0.1	0.4	0.01	3.9	95% CO_2 2% N	−100 (in winter)	yes
Jupiter	800	320.	2.7	3,000.	1.3	H, He	30,000	yes, on moons
Saturn	1,400	95.	1.1	unknown	0.7	H, He	unknown	unknown
Uranus	2,900	15.	1.0	unknown	1.6	H, He	unknown	unknown
Neptune	4,500	17.	1.5	unknown	2.3	H, He	unknown	unknown
Pluto	6,000	0.1	unknown	unknown	unknown	unknown	unknown	unknown

locked in a tumultuous soup of colliding photons and particles. This mix continued for several hundred thousand years until the temperature lowered sufficiently to allow electrons to bind with atomic nuclei. But long before the separation of light from matter, the intense photon radiation was having its effect. Neutrons and protons were continuously colliding, building toward helium and heavier elements. Just as rapidly, the photon radiation was smashing these nucleides apart, back into free neutrons and protons. The moment that the temperature fell to the critical level at which free neutrons could form stable bonds with protons, the free neutrons joined with hydrogen and formed deuterium and then helium. This nuclear stability was reached at a temperature just under $10^9°K$. The composition of the universe at that time was 75 percent hydrogen and 25 percent helium. Except for the heavier elements formed by fusion in the cores of later stars, the composition remains as such today.

Nothing seems overly portentous in this sequence of events until one considers the consequences that variations in the binding stability of neutrons and protons would have on the development of matter. Suppose that instead of this binding being stable at $10^9°K$, the temperature of stability was 10^{10} or 10^8. Our universe would be a very different place. (Instead of changing the temperature of nuclear stability, we might equally as well have theorized that the starting temperature at the creation was different than it was and so the universe would have cooled to the $10^9°K$ temperature earlier or later. This would avoid arguments against changing values for one of the inherent forces of matter.)

If $10^{10}°K$ was the temperature of stability, then stability would have been reached at approximately one second after the Big Bang. The composition of the universe at that time was approximately 25 percent neutrons and 75 percent protons and the particle plus energy density of the universe was some 400,000 times that of water. This high density would have caused rapid fusion among particles and therefore rapid building of heavier nuclei.

How would this affect us? First of all, we probably would not be here. Immediately, the composition of the universe would have shifted from its present 75 percent hydrogen and 25 percent helium to a 50–50 ratio as the abundant free neutrons joined with protons to form helium. The high particle density would have changed more, perhaps all, of the hydrogen into nuclei of heavier elements. Little or no hydrogen would have remained. No hydrogen means no significant solar radiation. The stellar furnaces would not have burned as they do today, because the energy of stars is fueled almost entirely by the fusion of hydrogen into helium. Those elements heavier than helium, which life now gleans from the residues of supernovas, would have been abundant. But the hot spots of the universe, which we call stars, would not be there to provide life-giving energy.

Had nuclear stability been delayed until the universal temperature cooled to 10^8°K, then instead of having an abundance of heavier elements and a dearth of hydrogen, as we saw in the previous scenario, there would be hydrogen and not much else in the universe. Approximately 300 minutes had elapsed before the expansion of the universe had lowered the temperature to 10^8°K. Although neutrons bound in a nucleus are stable and do not decay radioactively, free neutrons are radioactive. They decay with a 15-minute half-period. The 300 minutes that elapsed before reaching 10^8°K would have allowed almost complete decay of all free neutrons. Nuclear synthesis requires neutrons. A universe with no neutrons means a universe composed of hydrogen and no other elements. There is no place for life in such a universe.

The inflationary epoch fine-tuned the distribution of matter in our universe, thus allowing it to expand over millennia. Then the nuclear force fine-tuned the nucleosynthesis of this matter so that it provided the needed hydrogen, the needed neutrons, and the needed elements for life. But that is not the end of the list of too-good-to-be-true events.

We see from the data in Table 4 that the well-directed nuclear synthesis produced carbon, the fourth element in the list of abundances, as the most plentiful element that is solid in the temperature range of liquid water. The first three most abundant elements are gases. It also produced hydrogen and oxygen, the atomic constituents from which water is made, as the first and third most abundant elements.[1]

Is this important? You bet your life it is!

Life is based on carbon and water. Without carbon and water, you and I and all the biosphere would not exist. All living organisms are approximately 80 percent water. If we remove the

TABLE 4. THE RELATIVE ABUNDANCES OF THE MOST COMMON ELEMENTS OF THE UNIVERSE

Element	Atomic Number	Relative Atomic Abundance in the Universe
Hydrogen	1	2.7×10^8
Helium	2	1.8×10^7
Oxygen	8	1.8×10^5
Carbon	6	1.1×10^5
Neon	10	2.6×10^4
Nitrogen	7	2.3×10^4
Magnesium	12	1.1×10^4
Silicon	14	$1.0 \times 10^4 \; (= 10,000)$
Fluorine	9	9000
Sulfur	16	5000
Argon	18	1000
Aluminum	13	850
Calcium	20	625
Sodium	11	600

water from our bodies, half of the remaining dry weight is carbon. There is good reason for this. Water, in its liquid state, serves as a universal mediator in both simple and complex chemical reactions. Carbon is the elemental jack-of-all-trades. It falls halfway between being a metal and being a nonmetal. That means it can form compounds with most other elements. Most important, carbon can form long chains with itself and include in these chains branches and rings. From such chains, often hundreds of carbon atoms long, the structure of all life is built.

Had nucleosynthesis skipped over carbon and gone on to the next most abundant element with metal and nonmetal properties, silicon, the existence of life would be questionable. As a chemical intermediary between metal and nonmetal, silicon, like carbon, forms a variety of compounds. It dominates the compounds of the mineral world. But unlike carbon, silicon lacks the crucial life-giving ability to form long chains. Molecules containing more than two or three silicon atoms are rare and so the long molecules of life such as DNA and proteins could not be based on silicon chains. Unlike many carbon compounds, silicon compounds are almost all solids in the temperature range that water is liquid. So the water-based reactions of life would not be possible in a silicon-based system.

The result of these differences is evident in the biosphere. Although the cosmic abundance of silicon is almost 10 percent that of carbon, there are no known life forms based on silicon. Carbon fits the bill as no other element can.

As our awareness of all these fortuitous circumstances in our physical world increases, we can see that we almost might not have made the scene.

Stars in their early stages are observed to lose vast amounts of matter in one explosive burst. This phenomenon is known as the T-Tauri phase of a star. The Sun was probably no exception. The T-Tauri phase of the Sun produced a solar wind so fierce that it literally blew all residual interplanetary gases into outer space. But the devastation touched more than just the interplanetary matter. Atmospheres of the planets were whisked away as well. Yet here we are breathing a life-supporting mixture of 20 percent oxygen and 80 percent nitrogen.

The timing of the sudden and violent solar wind must have been most propitious. The T-Tauri wind, which blew out a mass equal to two or three present-day Suns, occurred not too early and not too late. Too early would be before aggregation of matter into planets. Had that happened, the frozen gases, held as frost on the surfaces of the dust and rock particles, which were to cluster and form the planets, would have evaporated and been lost to space. The result: an Earth without water. Oxygen probably would have remained. Evidence indicates that it joined the Earth held as solid metallic oxides. But life without water is not likely.

Just as there is a "too early" for the T-Tauri wind, so also there is a "too late." A T-Tauri wind that comes after the formation of the planets would seem less threatening. Most of the water and other volatiles would already have been buried within the planets' mass and so would have been protected from the force of the T-Tauri wind. But this security is an illusion.

The source of our oceans and atmosphere are, to a large part, the gases and vapors released during volcanic eruptions. Because the T-Tauri of the Sun occurred *after* the agglomeration of the planetary matter but *before* the major volcanic outgassing that accompanied the melting of the young Earth, the water and other volatiles held within the Earth were protected from this solar wind. Had the Earth's meltdown occurred prior to the Sun's T-Tauri phase, the Earth's volatiles would have been exposed on the surface of the Earth. The devastating force of this wind would have swept the water and gases of the biosphere out to space. If this had happened, our planet would look much like Mars, probably a lifeless landscape with small amounts of surface water and only traces of water vapor in an atmosphere so thin that its surface pressure is a mere 1 percent of the Earth's.

The surface of Mars shows clear evidence of a bygone era when erosion caused by flowing water was extensive.[2] This phase of Martian history is recorded in the form of serpentine channels carved in the now-arid land. Photos of these channels could easily be mistaken for pictures of dried riverbeds in the southwestern United States. There is the possibility that some of the missing Martian water is now stored underground. Certainly on Earth vast reservoirs of water exist underground. If these also exist on Mars, then there is the potential for "subterranean" life within, if not on, the dusty Martian planet. We find that the Earth lucked out with just the correct mass in just the correct location.

Our average distance from the Sun is approximately 150 million kilometers. Venus is only 30 percent closer than we are but the difference is crucial. Surface temperatures on Venus are 500°C. At this temperature, zinc and lead melt. Wood burns spontaneously and glass is almost a soft putty. Not much chance for life under these conditions!

The annual variation in distance from the Earth to the Sun is only 4.5 million kilometers, that is, only 3 percent of the total distance. (We are the 3 percent closer in January.) This small annual variation means that the Earth's orbit is almost circular. The orbit of Mars, the planet just next door, is quite elliptical, causing a variance in the distance to the Sun of 50 million kilometers during the year. If the Earth had this variance, our crust would deep-fry each January. In fact, if our distance from the Sun were only 10 million kilometers less, that is, a change of less than 7 percent, calculations indicate that the increased solar heat would prevent water vapor from condensing. There would be neither rain nor oceans. We may all complain about the weather, but in reality we have much to be thankful for.

For life as we know it, water in its liquid state is essential. On the Earth's surface, water is a liquid between 0°C (32°F) and 100°C (212°F). That temperature span, although it represents the range of human sensations from bone-chilling cold to scorching heat, is less than 2 percent of the total temperature range measured in our solar system. There is good reason that human exposure to this relatively narrow temperature range is bracketed by sensations of discomfort. Our nervous system is telling us that life cannot exist outside these not-so-extreme extremes.

The positioning of the Earth relative to the Sun almost presents a cosmic double bind. The Earth needs to be close enough to get the warmth of the Sun for life. However, accompanying the warming rays of sunlight are heavy doses of solar ultraviolet (uv) radiation. Uv radiation is so lethal that it is used today to sterilize countertops. At a distance from the Sun where temperatures are acceptable to life, the intensity of uv radiation wreaks destruction on organisms not protected from it. Life bathed in uv light would not last long.

The Earth's early atmosphere was produced by volcanic gases. It was rich in carbon dioxide. In time, plant life appeared and through photosynthesis contributed the oxygen we breathe to the atmosphere. Upon striking oxygen, uv radiation forms ozone, and ozone has the wonderful characteristic of absorbing uv radiation while being transparent to visible light. With an ozone layer in the upper atmosphere, life at the Earth's surface is shielded from the lethal uv radiation while the nonlethal sunlight is able to penetrate to warm our planet and provide energy for photosynthesis.

Uv radiation is able to penetrate only a few millimeters of water. This gave life a chance to develop within the oceans prior to the presence of oxygen in the atmosphere. Oxygen produced by aquatic algae formed the initial ozone screen. Life could then emerge from its protected berth in water and populate the land.

The fine balance of the Earth's composition is made manifest by another characteristic of the planet we call home: its radioactivity. The young Earth contained enough radioactivity to have heated and melted during its early development. Evidence of this internal heating is the increasing temperature experienced as we dig into the Earth. By the time we are 1 kilometer within the Earth, as in the diamond and gold mines of southern Africa, air-conditioning is required to make the mines habitable. Compared to the total radius of the Earth, these mines penetrate just the narrowest film of the Earth's surface. Three hundred kilometers below the surface, the temperature exceeds 1500°C and at our planet's center the temperature reaches 4200°C.

The heat that melted the young Earth came primarily from two sources: the kinetic energy as particles fell together forming the Earth and the decay of radioactive nucleides contained within these particles. The kinetic energy was a one-time contribution to our thermal balance. Radioactivity is a continuing, but ever-decreasing, phenomenon.

Because of the decrease in radioactivity over time, the Earth now has a solid crust. The fact that the core of the Earth remains molten is stamped indelibly on our daily lives.

The motion of the molten iron mass within the Earth's core produces the magnetic field with which we are familiar. This magnetic field does much more than merely allow us to set a compass. Its force diverts much of the potentially lethal cosmic radiation that reaches the vicinity of Earth. We live under a literal magnetic umbrella. Were this cosmic radiation not deflected, it would bathe the surface of the Earth with a continual shower of life-devastating ionization. Ironically, there is no positive explanation for the origin of the Earth's magnetic field. The continual flow of the iron core maintains the magnetism, but the cause that started the magnetism has eluded researchers.

Understanding the sources of our water is a lesson in geophysics. For decades following the realization that the Earth was at one time molten, it had been assumed that the water of the Earth's oceans came from the condensation of steam held in a primeval cloud that surrounded the molten Earth. The geophysical constraints on the size of such a cloud are that the upper surface of the cloud had to be hot enough to maintain the water as a vapor and close enough to the Earth's surface so that the gravitational force would keep the water vapor from escaping to space. These requirements limit the maximum possible size of the cloud. Calculations made in the early 1950s showed that the amount of water able to be held in such a cloud could account for only a small fraction of the 1,400,000,000 cubic kilometers of water contained in today's oceans.[3] A different source for the Earth's waters was sought.

The biblical version of the origin of the Earth's waters does not include a primeval cloud. It is written, "And steam would go up from the Earth and water all the face of the ground" (Gen. 2:6). The Bible stated that somehow the Earth was a source of steam that on rising would condense and water the ground.

In 1931, R. W. Garanson had found that water dissolves readily in molten silicate, basaltic, and granitic rock. These are large constituents of volcanic lava. Molten rock can contain approximately 5 percent of its weight as dissolved water. As the rock solidifies, however, the water is expelled as steam. The quantity of steam from the volcanoes of a cooling Earth can account for the volume of water in our oceans. Chemical similarities between volcanic steam and ocean water indicate that this steam, rising from a cooling Earth, was very likely the source of our oceans. Here truly, the truth of Genesis 2:6 sprang from the Earth.

But if a little volcanic activity is good, a lot is not better. The internal heating of the Earth produced an amount of volcanic activity that was enough to liberate the water needed for oceans, the gases needed for the atmosphere, the molten iron core needed for our protective magnetic field, while being insufficient to make volcanoes and earthquakes a continual occurrence. The result of excessive volcanic activity would be a dusty and dark atmosphere, perhaps of the transitory type thought to have caused the dinosaur extinction.

It takes very special conditions to keep life going. The location, size, and composition of the Earth are nearly ideal for this task. The terrestrial suitability for life is not, however, proof for a Divine plan or Divine intervention. There are some hundred billion stars in our Milky Way galaxy and possibly a hundred billion galaxies in the universe. Each has billions of stars. Planet formation is a likely phenomenon. Nine formed in our solar system alone.

Within the vast reaches of the universe, there may be many planets suited for life. Whether they *do* have life is a different matter. Random chemical reactions are just as unlikely to produce life elsewhere as they are unlikely to have produced life here on Earth. But if primitive life did appear on other planets circling other stars, the physical laws of the universe—the nuclear, electromagnetic, and gravitational forces that became inherent parts of all matter at the creation—could, in theory, guide this young life from a simple state at its inception through to the formation of the complex creatures we have on that most suitable of homes for life: the Earth. This can all be explained by theoretical, emotion-free equations in a physics textbook.

What eludes us all is the physics that can bridge the immense chasm separating the most complex inorganic molecule from the orchestration that characterizes even the simplest living form.

Notes

1. Data in Table 4 are calculated from data presented in Dickerson, "Chemical evolution and the origin of life," in Dobzhansky et al., *Evolution*.
2. Schwartz, "Muddy evidence," *Scientific American* 260 (June 1989):13.
3. Von Arx, *An Introduction to Physical Oceanography*.

CHAPTER 9

Mankind: Body Plus Soul

■■■■■■■■■■■■■■■■■■■■■■

Biologists, impressed by the inherent improbability of every single step that led to the evolution of man, consider what Simpson called "the prevalence of humanoids" exceedingly improbable.

—Ernst Mayr

We have seen that the almost immediate appearance of life on the newly formed Earth is so highly improbable that it must be removed from the category of an inevitable event occurring among random chemical reactions. The appearance of life can only be attributed to a series of interactions guided by phenomena, natural or Divine, which have yet to be discovered by scientific inquiry. We have learned that the newly formed life required such precise physical conditions that only a finely tuned universe could possibly satisfy them. We must now address the problem of progressing from the simple forms of microbial life recorded as 3.5 billion-year-old fossils to the complexity of modern mankind.

Paleontologists and theologians *agree* that the record of life on Earth reveals a sequence in which the date when an organism appears in both the fossil record and the biblical account corresponds to the complexity of that organism. This progression does not necessarily mean that there was a smooth or continuous development from bacteria to mankind. Paleontologists and theologians also agree that the evidence we have at hand, be it biblical or scientific, is for the punctuated formation of life, that is, the appearance of new species with no direct connection between the new species and the species that predated them.[1] The so-called missing links are still missing! It is to this discontinuous record, so much in contradiction with the Darwinian concept that nature does not make jumps in the evolutionary process, that Niles Eldredge referred when he said, "The pattern we were told to find for the past 120 years *does not exist* [emphasis added]."[2]

The first chapter of Genesis is a clear description of an unfolding through time, not only of the inanimate objects of the universe, but also of the types and forms of life. It was not until the third day of Genesis that a dry Earth was available for populating (Gen. 1:9), and yet biblical scholars acknowledge that the matter from which all that ever was, is, or will be found in the universe was created in the beginning of the first day. This primordial stuff, we have learned, had to be shaped and reshaped into substances from which stars and planets and life could be drawn.

Those accepting evolution through natural selection and those believing in Divine guidance in our universe see plant life as the first form of life on Earth. The forms of this plant life are, however, quite different for the two schools of thought. The fossil record describes a microscopic form of flora, bacteria, and algae as the first living substances on Earth. The microfossils that revealed these most early life forms became part of the paleontologists' library only a few decades ago. Prior to the 1950s they were virtually unknown. It is this first life that required intimate guidance, Divine or otherwise, to come forth from inert matter. Molecular biologists of the caliber of Francis Crick and Leslie Orgel, scientists who are certainly familiar with the complexity of life, are unable to explain life's origins. It is they who have suggested nonearthly sources to explain the start of life on Earth.[3]

Because the Bible talks in the language of man, it consistently deals with aspects of life as they are directly sensible by humans. It is, for example, due to this human orientation that laws of biblically permitted and restricted foods (Lev. 11) discuss only forms of life visible to the unaided eye. Crustaceans are among the animals prohibited from consumption. Yet a glass of water or a leaf of lettuce need not be examined by microscope to learn if a microcrustacean might be present. Because it would have been unrealistic to expect a mass of newly freed slaves 3400 years ago at Sinai to grasp the meaning of bacteria and microalgae, it makes sense that the biblical record of life starts with that which is visible.

At the command of the Creator, the Earth itself brought forth grass and herbs as the first life described in Genesis. The appearance of life in the biblical record is marked by the absence of a singularly significant word: *barah*, meaning "created." The potential for this greening of the Earth appears to have already existed within the Earth. As Nahmanides states when commenting on the origins of plant life, "God decreed that there be among the potentials of the earth a force which causes vegetative growth and bears seed."[4] All that was needed was the Creator's suggestion to activate this potential.

On the third day of Genesis, plant life appeared. This occurred just after the Hebrew term for water took on its present meaning. Here, in Genesis 1:10, it is described as the substance that fills the seas. Prior to this time, the term referred to the primordial substance from which all matter of the universe was to be formed. Because it was only on the fourth day that luminaries appeared in the firmament of heaven (Gen. 1:14), the presence of plant life on the third day might seem out of order. Light is one of the prerequisites for photosynthetic growth of plants.

Resolution of this seeming conflict is found in the use of the word *luminaries* rather than *light* in Genesis 1:14. Prior to the appearance of abundant plant life, the Earth's atmosphere was probably clouded with vapors of the primeval atmosphere. This would be in accord with information relayed from Soviet and U.S. spacecraft investigating the cloudy atmosphere of Venus. There was light on the third day, in the sense that the atmospheric vapors transmitted radiant energy. The atmosphere, however, was translucent, not transparent. Therefore, individual luminaries were not distinguishable. It was this diffuse light that provided energy for the initial plant life. Nahmanides states that the firmament, formed on the second day (Gen. 1:6), initially intercepted the light that existed from day one. He was not willing to comment concerning the composition of the firmament, because he considered it as one of the deep mysteries of the Bible.[5]

The early plant life actually helped clear the atmosphere through the process of photosynthesis, which removed carbon and nitrogen compounds from the atmosphere and incorporated them into cellular material. As these biologically driven reactions proceeded, the Sun, Moon, and stars, already visible in the firmament, became visible on Earth as individual sources of light. That Genesis 1:14–18 is describing this event from an earthly viewpoint is made clear by the reference to the Moon as a great luminary (Gen. 1:16). The Earth is the only celestial body close enough to the Moon to see the Moon as a great luminary.

A by-product of photosynthesis was and is the release of molecular oxygen. The accumulation of oxygen in the biosphere and the gradual conversion of the originally oxygen-free (anoxic) atmosphere to one that is rich in oxygen is just one of many examples, ancient and modern, in which life has altered the Earth's environment for its eventual benefit.

A record of the transition from a biosphere that was without oxygen to one rich in oxygen is found in a comparison between the biochemical pathways in advanced, nucleus-containing bacteria cells (eukaryotes) and the biochemical pathways of the more primitive bacteria, those without cellular nuclei (the prokaryotes).

Primitive bacteria, which also form the oldest of fossils, obtain the energy for the processes of life by fermentation of existing organic molecules. In fermentation, a process requiring no molecular oxygen, a large organic molecule is cleaved into smaller fragments. The increase in disorder (and entropy) that results from the cutting of the chemical bonds liberates energy. Part of this energy is used by the cell, but considerable amounts of potential chemical energy remain in the smaller molecules left as discarded fragments of the original large organic molecule. This biological process makes its impact on our lives each time we use a dairy product. Fermentation of the sugars and cellulose of a cow's feed in its oxygen-free rumen results in the formation of milk fat.

The advanced, nucleus-containing bacteria cells, all of which live in oxygen-rich environments, obtain their energy from a two-step process we refer to as respiration. The first step is an exact repetition of the fermentation process. But with respiration, the originally discarded molecular fragments are biologically combusted with the oxygen of the environment. As the term *combustion* implies, the food is oxidized, ultimately producing water and carbon dioxide. This combustion is so efficient that it gives the cell almost 20 times more energy than was obtained by fermentation of the same food. The fermentation process was developed during the time when the biosphere was still without oxygen.

With the advent of photosynthetic life and the subsequent availability of oxygen, respiration of organic matter became possible. The respiration process was then annexed onto the existing fermentation pathway of the cells. It was so efficient that it became the dominant metabolic pathway for life. What is important to note here is that the older fermentation process was retained and the new, more efficient respiration process was added onto the already existing mechanism rather than starting a totally new process with no relation to the old.

The appearance of oxygen at a concentration of 1 percent or more within the Earth's biosphere has left its mark on the geologic record. Molecular oxygen is a highly reactive chemical. As such, it induced oxidation reactions that previously were absent. The presence of the oxygen-rich compounds of uranium in sediments less than 2 billion years old and not in older sediments implies that free oxygen became available at about 2 billion years ago.

Iron in marine deposits provides another clue to the oxygenation of the atmosphere. The oxidized form of iron first appears as the rusty rock strata that account for much of the world's minable iron reserves. Radioactive dating of these iron deposits indicates that their formation and burial occurred during a span of several hundred million years slightly over 2 billion years ago. This is just (in geological terms) after the earliest fossil evidence of what appears to have been an abundance of oxygen-producing blue-green algae.

The ability of photosynthesis to oxygenate the world should not be underestimated. The current rate of photosynthesis can produce in approximately 2000 years all the oxygen present in the Earth's atmosphere.

The Bible records that animal life appeared in the waters on day five and on the dry earth on day six. This is after the Sun became visible in the firmament of heaven on day four. This timing is logical if the appearance of the Sun and other luminaries is in fact related to the oxygenation of the atmosphere. It was the availability of oxygen that allowed the development of life forms larger than bacteria and algae and also produced the uv-absorbing ozone layer allowing the population of dry land.

Because the use of food by fermentation is so inefficient, large amounts of food must be consumed to obtain the energy needed to support the organism's life. Only microscopic life can be sustained in this way. The large ratio of surface area to body volume usually inherent in small objects, such as bacteria, is essential for absorption of the large amounts of needed foods and the excretion of the copious amounts of waste products.

In sharp contrast to the low rate of energy production through food fermentation, the energy released by the use of foods through the respiration process is so great that the consumed food is sufficient to supply the energy requirements of large, multicellular plants and animals. Only after the biosphere became oxygen rich could the large forms of life develop. They appear in the fossil record just following the appearance of the geological evidence for an oxygen-rich atmosphere. In accord with this, both the fossil record and the biblical account place the appearance of large animal life after the appearance of oxygen-producing plants.

Photosynthetic growth gets its energy from light rays. It is so efficient that it produces large amounts of organic matter, which provide a renewable source of food for animals. With the advent of photosynthesis, life passed two barriers: the continued production of foods was assured, and the oxygen to use this food efficiently was provided.

THE PERSISTENCE OF THE MISSING LINK IN THE FOSSIL RECORD

But what of the fossil record on which I have relied so heavily to outline the appearance and flow of life on Earth and which is used so frequently to repudiate the validity of Genesis? Does it provide a satisfactory explanation of life's origins and development? A study of its details reveals that it is no more satisfactory than George Wald's erroneous assumption that random processes produced life on Earth. Wald was 100 percent correct in his thesis. Given enough time random chemical reactions would lead to life. Wald assumed that the needed time would be found. But it was not. Life appeared almost immediately on the newly formed Earth. Those people who assume that the fossil record provides proof for the theory of evolution through natural selection have fallen into the same trap as did Wald.

The fossil record of the mid-1800s, the time of Darwin's *On the Origin of Species*, indeed contained organisms ranging from the primitive to the complex. But there was no continuity within this record. With its gaps in fossil evidence, it did *not* demonstrate an evolutionary flow from the primitive to the complex. Darwin realized this and acknowledged the deficiency. Darwin also qualified the meaning of the struggle that led to survival. But as so often happens within a movement, his followers were more certain than he of his theory's truth. It was obvious to them that evolution *must* have occurred. They were certain that with more fossil evidence, the glaring gaps in the record would be filled. The steplike increases in complexity of life that the fossil record showed would give way to a smooth trend, a life curve as it were, leading from bacteria to trees and mankind.

But the intense paleontological efforts of the past hundred years have not produced the evidence. The fossil record of the late 1900s contains a hundred times more information than the record as it existed at the time when Darwin (who on learning that Alfred Russel Wallace had also developed the concept of evolution through natural selection and was planning to publish it) published his *On the Origin of Species*. Yet today's fossil record is as discontinuous as that of Darwin's (and Wallace's) time. In the past, the discontinuities have been brushed aside as temporary phenomena, problems that will disappear as our store of paleontological information increases. This is no longer possible. As Dr. Niles Eldredge of the American Museum of Natural History in New York stated so definitively, "The pattern [in the fossil record] that we were told to find for the past one hundred and twenty years does not exist."[6]

In regard to Wald's arguments, the finds of early microfossils essentially disprove the possibility of randomness as the answer for life on Earth. In regard to the concept of a gradual evolution in the forms of life, the expanding fossil library has shown that while the theory of evolution, defined as the natural selection of those forms of life best adapted to their environment, is excellent as a principle for organizing into systematic groups the various current and past morphologies, the theory of gradual evolution is *unsubstantiated* by the fossil record. Gradual evolution is a fundamental tenet of Darwin's theory. But there is no rhythmic flow from the simple to the complex. Staccato is a more accurate description of the record. Most serious paleontologists now accept that a form of punctuated evolution is the best that can be derived from the information that fossils present to us. A life form appears. There may be changes within the form, but its basic structure remains until it disappears and a new, different structure arises in its place, *suddenly*.

A gradual, or slow, evolution of a new species from one that predates it, is *never* seen in the fossil record. In fact, an account of the fossil record shows that as far as macroevolution is concerned, *stasis, not change*, is the trend with all species yet formed.[7]

Some scientists cite the horse as an example in which fossils show a gradual evolution. However, a careful study of the line, from Eohippus to Equus (today's horse), reveals an erratic path, in which "some of the variants were smaller than Eohippus, not larger."[8]

This lack of paleontological evidence has not diminished the enthusiasm of paleobiologists for evolution via natural selection. The information missing in the fossil record is now sought by inference from similarities in genetic and metabolic processes among organisms of different species.

The number of chromosomes might have shown a positive correlation with increasing complexity of the organism, and increasing complexity might imply an advance in the new form over that which preceded it. Corn has 20 chromosomes, the mouse has 40, and humans have 46. Unfortunately for this theory, the potato has 48. (Considering the frequently deleterious state of human affairs, the correlation may be valid!)

Relatedness among different forms of life has been sought using the similarity, or lack of similarity, in proteins found in their bodies. Because the proteins of an organism are the product of the genetic coding of that organism, similarities in the proteins imply similarities in the genetic heritage. Some specific proteins are common to organisms as dissimilar as bacteria and wheat and man. As we discussed previously, the likelihood of a chance production of two nearly identical proteins by two separate organisms is less than one in 10^{130}, or approximately zero. The similarity of the proteins in diverse life forms is statistically strong evidence that they have a common ancestor. It is counterproductive for students of biblical tradition to ignore the results of this research.

The strongest evidence for a common ancestor of all life is the similarity of genetic material among all forms of life. DNA and RNA are so complex that the probability of their independent development in the plant, fungal, and animal kingdoms is vanishingly small.

Having a single primal source of life poses no more of a problem to the theologian than it does to the paleobiologist. The composition of the Earth contained the ability to "bring forth" life (Gen. 1:11, 24). The Talmud describes a swamplike interface between earth and water as the place of the origin of animal life.[9] The material aspects of our bodies are so intimately connected with the Earth that the biblical name chosen for the first member of mankind was Adam. *Adam* in Hebrew means "earth, or soil" (*adamah* in Hebrew). We are told explicitly that our physical origin is indeed the Earth (Gen. 2:5).

Like the fossil record, the biblical description of life's unfolding is punctuated, not smooth. However, between theologians and Darwinian paleontologists, there is a dramatic difference in how these punctuations are interpreted. Biblically, the transition from the nonliving to the living and the appearance of plants and animals are all marked by statements of God. The conditions of the environment may have had a role in determining which species would be able to flourish and which would perish, but the theological understanding is that the forms of life available to compete for survival in the changing world had teleological direction. If there is within the material of the universe an orthogenetic property, this orthogenesis is the result of a Divine plan as revealed in the Bible.

A divinely inspired teleology, or purposeful goal, is clearly at odds with the theory of random natural selection among species, the theory proposed by Darwin. But during the past few decades, modifications in Darwinian theory have brought natural selection considerably closer to the biblical theory. The most dramatic shift in opinion has been the acceptance that *immediate* random mutations are probably *not* the source of changes within a species.

When the characteristics of an environment are altered, the populations within that environment undergo stress; those well suited to survive the new stress thrive. But what are the sources of these variant traits? Based on the advances in molecular genetics during the past thirty years, it is now believed that the adaptive variant was already contained within the genetic material of the parent generations. It had been expressed by a small number of the population's members in past generations. When the environment demanded this trait for survival, the number of individuals with this trait increased greatly. This concept, studied by population geneticists, is the mainstream theory of every aspect in current evolutionary research. The range of genetic variation able to be contained within a given species is evident from the fact that among the 4 billion representatives of the human race, no two (excluding those originating from the same fertilized egg) have been found to be identical.

It is this variety, common to virtually all life, that is the bane of users of insecticides. The rapid adaptation of the housefly to DDT is not the result of chance mutations of their genetic material producing, just in the nick of time, the needed resistance to the chemical. In the pre-DDT world, the resistance was already present, but few flies expressed the genetic trait. When the introduction of DDT into the flies' environment demanded this trait, those progeny in which it was expressed became the more numerous strain. The resistance of hundreds of insect species to chemicals that were toxic to their ancestral generations is noisome testimony to this ability for change to occur within a morphotype.[10]

The fossil record also provides evidence for such changes but only if we work *within* a single life form or species and do not attempt to impose a preconceived notion of what the "missing link" should be that might join different morphotypes. An example of this is found within the 2-million-year record of the animal genus *Homo*, where variations in skeletal size are apparent. Cranial capacity (an indication of brain size) increased during this period from approximately 900 cubic centimeters in volume in *Homo habilis* to 1400 cubic centimeters in Neanderthal man. It is about this value in modern humans as well.[11]

In the Bible, changes in characteristics of humans, the only living being to be considered in detail, are also discussed. The most obvious of these changes is that of life span. In the generations following Adam, ages exceeding 900 years are reported. Matushelah reached the age of 969 years (Gen. 5:27). By the time of David, "The days of our years are 70 or by reason of strength 80 years" (Ps. 90:10). That is just the extent of old age that we enjoy today. The average life expectancy at birth in the developed world is now approximately 71 years. (It is approximately 45 years in Africa and 55 years in Asia.) What caused the drastic decline in longevity from the time of Matushelah? Nahmanides (on Gen. 5:4) attributed the reduction to physiological stress resulting from lasting changes in the climate and atmosphere during and following the flood in Noah's time.[12] He discounts the suggestion that lifestyle or diet could bring about such a drastic reduction in longevity.

What we learn from this is that it is in accord with biblical tradition for changes to occur within a given species and that these changes can be the result of environmental influences. Natural forces influence the development of life, but tradition insists that these forces at key junctures were and are divinely directed. It is the "And God said . . ." that signifies the imposition of the Divine will, punctuating the natural processes of the world during the six days of genesis.

The rub in the controversy between evolutionists and theologians is that, objectively, the fossil record also presents a description of life's progression as following a path of punctuated development. Abrupt changes in forms of organisms, as life moves from the simple to the complex, are the rule and not the exception. A line of animals appears, perhaps attains dominance, and then disappears from the record to be replaced by another form. In some instances, these punctuations are associated with mass extinctions of whole groups of species.

There have been some five or ten major extinctions since the wide variety of life forms that marked the beginning of the Cambrian geologic era some 600 million years ago. Some have been associated with changes in the pattern of landmasses. The breakup of the early supercontinent, to which most of the Earth's landmass was attached, 580 million years ago, produced wide expanses of shallow seas along the newly formed continental shelves. Marine life flourished in what seems to have been the shoals of these sun-warmed waters. Extensive deposits of marine fossils, dated at this period, attest to the fertility of the waters. Then, 250 million years ago, the landmasses reassembled and formed the supercontinent designated Pangaea. The shallow seas disappeared as the landmasses pressed together. The result: 90 percent of the marine species present prior to this assemblage of land disappeared from the fossil record. Reptiles became the dominant animal, but a limited ecological niche was also found on this new continent for the first of the mammals.

Perhaps the most celebrated mass extinction is that of the dinosaurs. The dinosaurs made their appearance 220 million years ago and became the dominant form of reptile. They maintained sway over the animal kingdom for 150 million years. Judging from the fossil record, but without the wisdom of hindsight, they were the form of the future. They were here to stay. And it is not clear that they were dull-witted, cold-blooded reptiles. Based on comparative anatomy, including study of the shape and structure of their bones, Professor James Valentine of the University of California has suggested the possibility that dinosaurs were both warm-blooded and lively.[13] During the 150 million years of their presence, the fossil record shows that there were five extreme and three lesser mass extinctions. Dinosaurs survived them all. Finally, 65 million years ago, what can be described as a cosmic force de frappé did them in, permanently, along with all other animals larger than about 10 kilograms. The mammals, which had been kept in their diminutive place during the dinosaurs' reign, made it through whatever it was that forced out their competitors. In the newly opened ecological space, that special form of mammal, the primate, flourished.

What was it that caused the extinction 65 million years ago, and possibly several of the previous extinctions? The account of what is now generally thought to be the causing factor reads like a rerun of Immanuel Velikovsky's Worlds in Collision.

In the 1950s Velikovsky proposed that a continuing series of encounters between the Earth and extraterrestrial bodies, such as comets, had seeded life on Earth and then influenced the development of this life. The impact of these interactions extended even to the sociological path presented in the Bible. The thought of such a significant extraterrestrial role in our affairs was soundly quashed by the scientific establishment. But it is just such a cosmic encounter that is now being called on by respected scientists to explain the extinction of the dinosaurs and other extinctions as well.

In geographically separated regions, Walter Alvarez and his Nobel Prize–winning father, Luis, have found traces of a rare metal in sedimentary layers of rock dated at the time of the extinction of the dinosaurs. The presence of this metal, iridium, is associated with certain classes of meteorites. The concentrations of iridium found by the Alvarez team are 30 times greater than normally present in sedimentary rock. This finding has been replicated at locations around the world. It implies an extraneous input to the global environment. A collision with a meteor approximately 10 kilometers in diameter could have supplied this iridium. The aftermath of the impact would have had devastating effects on the biosphere. Quadrillions of tons of dust would have been blasted into the stratosphere reducing the amount of sunlight reaching the Earth's surface to an intensity less than that of a quarter moon. The Earth would have experienced, in effect, an extended nuclear-type winter. Photosynthetic rates and temperatures would have plummeted. In the altered ecology of the Earth, only the most resourceful of species could have avoided the extinction suffered by the dinosaurs.[14] The data for this scenario are sufficiently plausible to have warranted publication in the highly respected journal *Science*.[15]

Similarly caused mass extinctions are now suggested to explain many of the gaps and steplike changes of life forms found in the fossil record. The dinosaur extinction is particularly pertinent for mankind. Mammals coexisted with dinosaurs for the 150 million years that dinosaurs dominated the animal kingdom. But the mammals occupied a diminutive niche. It was only following the dinosaur extinction that mammals and especially primates were able to extend their domain.

Was it a time of guided evolution, filled with miracles we might easily misinterpret as fortuitous acts of nature? Hidden miracles are no less Divine than was the splitting of the Sea of Reeds, when the Israelites left Egypt. There, also, no direct proof of Divine intervention was given. A force of nature, a strong east wind that blew all night (Exod. 14:21) caused the sea to go back. The phenomenon was sufficiently natural for the Egyptians, now in hot pursuit of the Israelites, to mistake the cause as being totally natural and to follow the Israelites onto the momentarily dried seabed.

No one can verify that a meteor's collision with the Earth was a Divine intervention to direct the development of life toward mankind. Abrupt ecological changes, whether caused by colliding landmasses forming a Pangaea or by the tail of a comet brushing our atmosphere, might have been chance events that just happened at a propitious time. The difficulty in evaluating, objectively, the fossil record in relation to the theory of natural selection is that natural selection is so appealing as a means of grouping life into organized categories. But evidence for gradualism, the measured step-by-step selection of the fit over the unfit, is lacking in the record. Only by subjectively implanting a series of missing links among the fossil data or by accepting an evolution marked with abrupt punctuations can the advance of life be demonstrated by the fossil record. The latter option is, by necessity, now accepted by leading evolutionists. Has Velikovsky gained credence within the scientific community?![16]

READING THE FOSSIL RECORD

Subjectivity in "reading the data" seems to be an accepted trait among those who see evolution as a gradual process. It is this subjectivity, this adding of the wished-for-but-not-found data to the argument, that has allowed the proponents of natural selection to substantiate their claims and concurrently to degrade the credibility of the biblical tradition which tells of a punctuated development of life. A glaring example of this proving a point irrespective of the facts is contained in the writings of Ernst Mayr.

Mayr's credentials are impeccable: Ph.D. in zoology from the University of Berlin, former associate curator at the American Museum of Natural History, professor emeritus at Harvard University. When Ernst Mayr speaks (or writes) people listen! In a lead article in the prestigious journal *Scientific American* titled "Evolution," Mayr acknowledges that the "inherent improbability of every single step that led to the evolution of man [makes] the prevalence of humanoids exceedingly improbable."[17] But in his goal to prove that natural selection can account for mankind's existence, Mayr brushes aside the concept of a Divine teleology as "the creation myth of primitive people and of most religions."[18]

Now there is an aspect of Darwin's understanding of evolution that is in contradiction to the current understanding of changing morphology. Darwin believed in the inheritance of acquired traits. This process would mean that the favored use, or conversely the lack of use, of a body part would be expressed in the structure of that body part in the next generation. If you spend your time reaching for apples at the top of the tree, your children will have long arms. From observations, it is clear that the person who reaches for those apples may develop a slight elongation of his or her arm. But there is no evidence in organisms larger than microbes that succeeding generations will inherit an acquired trait. Mayr, who holds Darwin in high esteem, proposed a reason for Darwin's belief in the inheritance of acquired traits.

Regarding the concept of the inheritance of acquired traits, Mayr writes, "It was a universal belief, grounded firmly in folklore (one expression of which was the biblical story of Jacob and the division of striped and speckled livestock)."[19] Mayr interpreted this biblical episode (Gen. 30:31–43) as only a reader having a preconceived notion of the facts could so interpret it. The correct meaning of this episode is quite clear and can be grasped either in the original Hebrew text or in translation.

Jacob is asked by his father-in-law, Laban, to fix the wages he (Jacob) is to get for tending Laban's sheep. Jacob, having been repeatedly cheated by Laban in the past (including Laban's switching brides on the then-naive Jacob), answers cleverly: Remove all the speckled and spotted sheep and goats from your flock. Leave only the white animals. I will shepherd the all-white flock and every speckled or spotted animal born from this flock in the future shall be mine. This shall be my wage. The white animals shall be yours. Laban accepted the deal and that day (not giving Jacob any chance for monkey business) he removed the goats and sheep that were so marked. These he gave to his sons to tend at a distance of three days' journey from Jacob and the all-white flock.

At breeding time Jacob placed at the watering troughs fresh, partly peeled stakes or rods taken from poplar and other species of trees. The flocks conceived among the rods. Mayr's interpretation of the text is that the streaked and spotted appearance of these partly peeled rods increased the frequency of the birth of spotted offspring. But the Author of the Bible was wiser than Mayr credits. Let us follow the text accurately.

Jacob had tended Laban's flock for over 14 years prior to the deal. He knew well their breeding habits and had exploited this knowledge successfully in the past. Indeed, "For it was little which you (Laban) had before me (Jacob) and it has increased to plenty" (Gen. 30:30). Through 14 years of observation Jacob had come to realize that within the potential of the white animals was the ability to produce speckled and spotted progeny. Following the deal with Laban, generation by generation (Gen. 30:40–42), Jacob ensured that the stronger of the flock bred with those producing the marked offspring. The weaker of the flock he let breed at will. This is stated explicitly in Genesis 30:42. Nowhere is it stated that the color of the rods affected the color of the offspring. Rashi (on Gen. 30:38) explains the purpose of the rods for those readers not able to catch the subtlety: "When the female animal [bending forward to drink and therefore raising her hind quarters] saw the rods [reflected in the water], she was startled backwards, and the male [standing behind her] copulated with her."[20] Rashi presents a graphic, if slightly baudy, description of Jacob's plan. The effect of the rods was to increase the rate of copulation of the selected males with the selected females. Mayr might have agreed with this biblical account of ancient selective breeding.

The biblical text is clear. It is unfortunate that Mayr did not go beyond a simple reading of the text. One can only speculate on what he was and was not able to grasp correctly from the fossil record, which at times is quite recondite, an entire limb often having to be extrapolated from a single flake of bone.

There are fossils. These fossils can be arranged to show a progression from ancient, simple morphotypes to the complex, multicellular forms of today's plant, fungal, and animal kingdoms. Does this evidence mean that life's flow is all by chance; that man is merely the latest of the links in the chain? The existing data might lead to this conclusion. Scrutiny of the same information has led me to a different but partly enigmatic conclusion, a conclusion that at some stage requires a leap of faith.

Our world seems to alternate between one driven by seemingly natural causes and one that is guided by the will and occasional intervention of a being that we have labeled "God." This understanding has at times the quality of the uncontrollable sequences in which an object in a painting may suddenly reorient to become the background. The painter M. C. Escher was a master at creating paintings in which such reversals occur at the twinkling of an eye.

In spite of the "inherent improbability of life" of which Mayr wrote, life appeared almost immediately on the newly formed Earth. Even in its simple forms, the indications are that this first life contained the majestically complex genetic material that we have labeled DNA and RNA. Two and a half billion years were to pass before multicellular life entered the fossil record. The transition from single-celled to multicelled life required the advance from nucleus-free bacteria cells to cells wherein the genetic material is organized within the nucleus. Also required was the transition from an oxygen-free, energy-inefficient biosphere to the energy-efficient and uv-shielded oxygen-rich atmosphere we know today.

The advance from single-celled to multicelled organisms was a complex orchestration of events. But how much more complex was the transition from the nonliving to the living, from inert mineral compounds to the extraordinary complexity of DNA and RNA found in even the most simple bacterium! The transition from single-celled life to multicellular life took 2.5 billion years. The fossil record shows that the transition moving from the nonliving to the living was accomplished in one-tenth of that time. Could this have happened so rapidly if random, unguided events were the creating forces? Statistically, even logically, it is so highly improbable that the likelihood is close to zero.

The evidence of the fossil record implies that either life was brought to Earth or that it was manufactured here by nonconventional forces (a Superintelligence or Creator?) or chance has played a trick on the statistics of mankind. If chance has indeed played a trick, then we have no basis for using statistical theory to analyze data and all physics is based on statistics, on the premise that the probable occurs and the highly improbable does not occur.

Explanation of the early part of the fossil record points us, inescapably, to forces not indigenous to nature as we experience it today.

What of the intermediate fossil record? Is it explainable as simply survival of the fit, as an open competition with no influences pointing toward some goal? Let's review the record. The fortuitous nature of the sequence is as intriguing as its staccato pattern.

The Earth solidifies. Erosion begins reshaping the exposed landmasses. Subduction and compression of this eroded material forms the first sedimentary rocks. Within the earliest of these rocks, the first traces of life appear. Nucleus-free (prokaryotic) bacteria remain the sole representatives of life on Earth for 2.5 billion years. That time span is half of the total present age of the Earth! Landmasses on the move, 1.3 billion years ago, join to form the first supercontinent. Concurrently cells containing nuclei (eukaryotes) appear and rapidly become dominant. This becomes the status quo for another 600 million years. That, again, is half the time between the formation of that first supercontinent and today. The continent breaks apart. At the same time, the first multicellular life appears and rapidly becomes dominant. During the following 300 million years aquatic and land species radiate, filling the available ecological niches. Then 250 to 225 million years ago, again approximately half the time between the breakup of the supercontinent and today, the landmasses reassemble forming the new supercontinent Pangaea. Major extinctions, including approximately 90 percent of all marine species, occur. Shortly (in geologic terms) after this, dinosaurs and mammals appear but the race between mammals and dinosaurs for dominance is to the dinosaurs. Even the several partial extinctions that they suffer during their period of tenure do not allow mammals to become supreme. For 150 million years they rule our roost. Mammals, the size of mice and squirrels, may scurry among the dinosaurs' feet, but they grow no larger than that. Dinosaurs are the shape of the future. In the environment of their day, they defeat all comers. Then suddenly something (Something?) changes the rules of the game and a *force de frappé*, possibly a prolonged winter induced by a dust-clouded atmosphere that was in turn caused by the explosion of a massive meteor slamming into the Earth, destroys the dinosaurs. They disappear and primates, having appeared just prior to the demise of the dinosaurs, start their climb to dominance. It is as if this biologic order to which man belongs had been waiting in the wings to take up the drama of life. A succession of further environmental changes, especially the waxing and waning of glacial cover, occurs encouraging the primates' move from trees to the forest floor and then to open country where an upright stance is an advantage.

No one can prove that Divine inspiration was the source of the immediate appearance of life on the young Earth or that the abrupt changes in the environmental conditions, which we see in retrospect directed life's journey to produce animals in the shape of man, were directed by the Creator. What we can say is that the fossil record definitely does not show a journey ruled by chance, or prove an unhindered march in the survival of the fit.

CROSSING THE TIME LINE OF ADAM

Still this sidesteps what is clearly the bottom line in the debate between naturalists and theologians: What was the origin of mankind? There may be a dearth of fossil evidence describing the paths of early life, but a range of fossils exist showing that creatures almost identical in shape to humans have existed at least for the past million years and perhaps longer. We might be tempted to neglect the fragments of the early fossil record. After all, they themselves are not actually dated. Their ages are estimated from the age of the deposits at the location of the find. But there exist hundreds, even thousands, of fossils from the past 15,000 years. These are accurately dated by the carbon-14 concentration in the fossil itself. At the edge of the Arctic, the remains of whole villages of Eskimo-like dwellers have been found.[21] These existed 9,000 years ago. Among the finds, in addition to the skeletons, are relics of clothing and housing and the utensils of a socially organized daily life. Equally well documented pre-Adam settlements extend from France to the Ukraine. In these finds, in addition to the tools of daily life, there are finely fashioned sculptures of a multitude of animals, and more extraordinary, drawings that reveal the perception and hand-eye coordination required to show three-dimensional perspective in a two-dimensional representation. Our similarity to ancestral hominids extends to cranial capacity as well. It has not increased since the appearance of the Neanderthals, 100,000 years ago. In fact, there are Neanderthal fossils with cranial capacities reaching 1,400 cc. That is some 100 cc larger than today's humans and 300 cc larger than the brain of Anatole France, the acclaimed writer and winner of the Nobel Prize![22]

Three million years ago in a region of Hadar, Ethiopia, and in Laetoli, Tanzania, animals that walked upright with a smooth and erect bipedal stride first appeared. This means that for three million years a type of animal moved about with hands free for grasping and using objects. Yet during almost this entire period the manufacture of tools evolved only slightly, being confined to the making of stone cores and flint blades. Then quite suddenly, approximately 40,000 years ago, or to put this timing into perspective, at the start of the final 1 percent of the time that bipeds have been roaming the Earth, tools diversified at a rate far in excess of anything seen during the previous 99 percent of their tenure. Stone spearpoints, harpoons, bone needles, and statuettes appeared. This sudden abundance of material wealth plus a social structure that appears to have included care for the infirm (based on the remains of healed breaks in limbs, and diseased jaws which must have been toothless), indicates a new ability to organize the life of a community. The instrument of this organization, 40,000 years ago, was probably some form of speech.[23]

The shapes of the interiors of fossil skulls indicate that even the three-layered structure of our brains has existed for at least hundreds of thousands of years and it possibly extends back a million years. What appears to be the oldest part of our brain is a mass of nerves at the top of the spinal cord. In this region, referred to as the stem, or brain stem, automatic body functions such as breathing and heartbeat are controlled. Overlying the brain stem is the reptilian part of our brain, where the instincts for territorial control and for fight or flight are seated. These very instincts still strongly control the actions and reactions of today's reptiles. They are also part of our intellectual baggage. Above the reptilian brain lies the limbic system. With the appearance of this portion of the brain, mammals depart from all their ancestors. From there originate emotions, especially those related to the love and care for offspring. The most recent part of the brain to evolve is the cerebral cortex. This overlies the limbic system. It is from here that analytical thought and the ability for mathematics and language and forethought stem.[24]

The very fact that the brain is layered, with each successive advance in intellectual development literally placed on top of its predecessor, indicates a pattern in development of brain morphology that is similar to the development of other body functions. Recall that at the cellular level the processing of foods consists of the more recently developed oxic respiration cycle being appended to the older anoxic fermentation cycle. The fact that the development of a fertilized human egg proceeds through gill-bearing and tail-bearing stages shows that the ability to produce those organs is still within our genetic material. Our final structure is molded in part from a composite of these earlier forms. The flippers of the porpoise have retained the bone structure of the forearm and hand that forebearers of this mammal had. The gene site for this structure has been retained within the porpoise's genetic material just as humans retain the gene site for gills and tails.

In light of this additive nature of development, the sobering realization is that our brain is not *qualitatively* different from many lower animals'. While the cerebral cortex is largest in humans, and as such provides us with the ability to formulate language and conceptual reasoning, its structure is the same as it is in rodents, monkeys, and all mammals. For all, it is composed of vertical columns of interconnected nerve cells, each 0.2 to 0.3 millimeters in diameter. Each column forms a modular unit.

The add-on nature of this modular design has enabled the expansion in mental capacity without a need to change the basic internal structure of the brain and nervous system. The social breakthrough of language, spoken and written, was accomplished by humans and their nearly, but not quite, human ancestors through the quantitative increase in these modular units, numbering billions in humans, and not by a qualitative change in the functioning of the brain.

The record is all but irrefutable. Either we believe that the

Creator placed fossils in the Earth to test our faith in a literal understanding of the biblical account of the Creation (and I do not subscribe to such a concept), or we must acknowledge that a form of animal life that was very much like human life predated Adam and Eve.

This latter option is not in contradiction with the established tradition on which I am relying. Well within the scope of biblical tradition is the fact of a *directed* evolution of man, one that arises from the pristine matter of the universe.

THE SOUL OF LIFE

Biblically, Divine punctuation in the flow of life is seen in the original creation of the universe, in the creation of the ability for animal life to arise from this matter, and in the tenfold repetition of "and God said" recorded in the first chapter of Genesis. The fact that the making of man is the most intimately described event of that chapter in Genesis implies that mankind is the targeted goal of those punctuations.

When on the sixth day, in the Creator's space-time reference frame, God decided to make mankind, the Bible first states that God will *make* man in God's image and likeness (Gen. 1:26). In the following verse it is written, "God *created* mankind in his image, in the image of God He *created* him, male and female He *created* them." The verbs *make* and *create* are both used, and so, from these two verses, it appears that both making and creating were involved in the appearance of the first of mankind. Later (Gen. 2:7), it is explicitly stated that mankind is *formed* from a previously existing substance, in fact, the same substance used to form fowl and land animals (Gen. 2:19). However, a special ingredient not mentioned before is summoned at this juncture. God breathes a *neshamah*, a "soul of life," into this creature and man became a living being.

Nahmanides in his *Commentary on Genesis* and Maimonides in his *Guide for the Perplexed* both state with no equivocation: Every material thing that was eventually to exist was derived from what was created in the first instant of creation. That was the only material creation.[25] From that ethereal mass of pure energy and exquisitely thin substance, stones and galaxies and humans were to be formed. We are products of the Big Bang. We are, in fact, made of star dust. The material aspects of man are totally rooted in the universe.

The specialty of mankind is not the physical attributes we have. All primates have grasping upper limbs and overlapping binocular vision. Based on the position of the larynx inferred from skull shape, articulate speech has been possible for over 100,000 years. The size of the cranial cavity, from which brain size is estimated, has not changed much in the primate we call Homo sapiens for the past 100,000 years.

Because our physical makeup is not what makes us unique and because sages and scientists agree that the matter of mankind has a common origin with all other universal matter, a theological problem is not posed by having the physique of mankind develop through an evolutionary process. Indeed, Nahmanides comments on Genesis 1:26 that the "us" of "And God said let us make man" refers to joint contributions by God and the existing Earth. Here Nahmanides repeats that only on the first day was matter created from nothing.[26] Thereafter all things were formed from the existing elements. For this reason it is written that at God's command (punctuation), the waters and the land brought forth life.

However, mankind and his predecessors, although physically related, are not connected by a spiritual line of evolution. Homo sapiens roamed the Earth for some 300,000 years, in our space-time reference frame, prior to the appearance of mankind. The Neanderthals appear to have started burying their dead 100,000 years ago and their fossil remains as well as those of the more recent Cro-Magnon become increasingly similar in shape to human beings as the time before the present decreases. But neither the Neanderthal nor the Cro-Magnon evolved into human beings.

At a crucial junction some 5,700 years ago a quantum change occurred. This change is the reason for the biblically stated partnership between God and the Earth in creating mankind. Indeed, so intimate is mankind's connection with the Earth that the name chosen for the first of the species is Adam, which means "soil" in Hebrew.

All animals received a life-giving spirit, a nefesh in Hebrew. The animal that was about to become Adam was no exception. However, into the physical form that contained the nefesh of

Adam, the Creator placed an additional spirit, or soul, the *ne-shamah*. It is this that has set mankind apart from the other animals. "And the Lord formed man (*adam*) from the dust of the ground (*adamah*) and blew in his nostrils a soul of life (*neshamah*), and the man became a living being (*nefesh*)" (Gen. 2:7).

In *The Guide for the Perplexed*, Maimonides makes a remarkable comment. In the time of Adam, he writes, there coexisted animals that appeared as humans in shape and also in intelligence but lacked the "image" that makes man uniquely different from other animals, being as the "image" of God.[27]

Nahmanides (on Gen. 2:7) observes that mankind developed through three distinct stages. The material of Adam's body was initially in the form of inert matter (the dust of the earth, Gen. 2:7). In the first stage of growth, there was a force that produced growth, "like that in a plant." Then with further Divine input, man was able to move, first as the fish and then as the land animals.[28] Here Nahmanides, still commenting on Genesis 2:7, refers to Genesis 1:20 and 1:24. These two verses describe the sequential first appearances of aquatic life and then terrestrial animal life. Prior to attaining the unique attribute of mankind, Nahmanides continues, the animal that was to become man had both the physical structure *and* the power of perception of a human. Only when this was accomplished was the spirit of God, the *neshamah*, breathed into him. Nahmanides concludes this discussion with the observation that the grammatical construction of this verse (Gen. 2:7) indicates that reasoning, speech, and all the other capabilities of mankind, while not being a part of the spirit, are subject to the spirit that was given to mankind alone among all the animals. God's direct and newly created contribution of spirit came to man only after the material part was intact. This contribution had no physical attribute. This *neshamah*, placed in mankind by God, was the last act in the making and the creating of mankind.

While the sequential development of mankind described by Nahmanides follows an evolutionary trend, the brief description that he presents (as is his style when dealing with "secret" or "cabalistic" matters of the Torah) precludes using this as a guaranteed description of the guided evolution of life from inert matter to mankind.

The author of Genesis stated explicitly that mankind was made *and* created in the "image," as a "likeness" of God (Gen. 1:26). Does this mean that there exists a corporal or physical similarity between man and God? It is a basic principle of the Judeo-Christian tradition that the Creator is not corporal and that no bodily attributes apply to the Creator. This seeming contradiction between the incorporeality of God and making man in God's "image" is resolved by the root meaning of the Hebrew word for *image*, which modifies the word *likeness* in the biblical text (Gen. 1:26). That meaning is "shadow." Man is indeed intended to present a likeness of God. But the likeness is not physical. It is the projection of God's acts as they appear in this world, God's shadow as it were. We are obliged to emulate God as we perceive God's interactions projected within our world.

Based on the literal meaning of Genesis 2:7 and 2:19, man and land animals and birds were *formed*, and they were formed from the same substance, the ground or earth. There is, in the two verses, an important difference in the spelling of the verb *formed*. Although in both verses the tense and person of the verb are identical and the structure of both verses is the same, when describing the formation of mankind an extra grammatically superfluous Hebrew letter, *yōd*, is added to the word *formed*. *Yōd* is the first letter in the Hebrew name of God and is also used as an abbreviation for God's name. By the addition of this extra *yōd*, we have been told that in the forming of mankind, God touched mankind in a way that was unique.

And that is why the description of progress in civilization, as recorded in the Bible, starts with Adam. The eternal and the earthly have become intimately linked in mankind. From Adam and thereafter, the space and time frames of God and the Earth are the same.

> And God formed man of the dust from the ground and breathed into his nostrils the spirit of life
>
> (Gen. 2:7).

Notes

1. Eldredge and Gould, in *Models in Paleobiology*, ed. T. Schopf, pp. 82–115; and Gould and Eldredge, "Punctuated equilibria: the tempo and mode of evolution reconsidered," *Paleobiology* 3(1977):115.
2. *The New York Times*, November 4, 1980, p. C 3.
3. Crick and Orgel, "Directed panspermia," *Icarus* 19 (1973):341.
4. Nahmanides, *Commentary on the Torah*, Genesis 1:11.
5. Ibid., Genesis 1:6.
6. *The New York Times*, November 4, 1980, p. C 3.
7. Ridley, "Evolution and gaps in the fossil record," *Nature* 286 (1980): 444.
8. Taylor, *The Great Evolution Mystery*, p. 230.
9. *Babylonian Talmud*, Section Hullin 27b.
10. For a clear introduction to population genetics see: Ayala, "Mechanisms of evolution," *Scientific American* 239 (September 1978):56–69.
11. Weaver, "The search for our ancestors," *National Geographic* 168 (1985):560.
12. Nahmanides, *Commentary on the Torah*, Genesis 5:4.
13. Valentine, in Dobzhansky et. al., *Evolution*, p. 66.
14. Gore, "Extinctions: what caused the earth's great dyings?" *National Geographic* 175 (1989):662.
15. L. Alvarez, et al., "Extraterrestrial cause for the Cretaceous-Tertiary extinction," *Science* 208 (1980):1095.
16. Gould, *Hen's Teeth and Horse's Toes*.
17. Mayr, in Dobzhansky et al., *Evolution*, p. 8.
18. Ibid., p. 4.
19. Ibid.
20. Rashi, *Commentary on the Torah*, Genesis 30:38.
21. White, "Visual thinking in the Ice Age," *Scientific American* 261 (1989): 74; and Putnam, "The search for modern humans," *National Geographic* 174 (1988): 439.
22. Weaver, "The search for our ancestors," *National Geographic* 168 (1985): 560; Dubos, *Celebrations of Life*, p. 20; and Gould, *Ever Since Darwin*, p. 181.
23. Garrett, "Where did we come from?" *National Geographic* 174 (1988): 434.
24. Lampert and Branwell, eds., *The Brain*.
25. Nahmanides, *Commentary on the Torah*, Genesis 1:1, 1:2, 1:4, 1:26; and Maimonides, *The Guide for the Perplexed*, part 1, chapters 27 and 29.
26. Nahmanides, *Commentary on the Torah*, Genesis 1:26.
27. Maimonides, *The Guide for the Perplexed*, part 1, chapter 7.
28. Nahmanides, *Commentary on the Torah*, Genesis 2:7.

CHAPTER 10

Our Cohesive Universe
Finding a Cause
■■■■■■■■■■■■■■■■■■■■■■

The theistic description of our universe, although developed millennia before, and therefore in isolation from, current scientific opinions, closely matches the perceptions of current cosmology.

W ith the information we have surveyed in our trek through cosmology and the history of life, can we demonstrate the potential or lack of potential for the infinity of a Creator within the limited space of the creation? If so, we may have learned whether or not there is a purpose for life that transcends our immediate desires. Let's review the main events that mark our flow from the primordial substance of the Big Bang to the universe of today to determine

1. Whether these events form a set of related events with a causal explanation of their relatedness (that is, a cohesive set), or whether they form a group of events with no causal relation to tie them together (an arbitrary set).
2. Whether the uniqueness of the universe and the complexity of life make our existence so improbable that an explanation is required to account for this existence.
3. Whether the biblical view of these events is sufficiently similar to the scientific explanation of these same events so as to require an explanation for this similarity.

The first of the cosmic events we discussed actually preceded the Big Bang by a speck of time. This was the inflationary epoch of the universe, a period lasting, by current estimates, from 10^{-35} to 10^{-32} seconds after the beginning (the creation?). This one-time, never-to-be-repeated, superrapid expansion set the universe on a course that was targeted toward life. It was not the start of a predestined flow to man. A closer analogy would be a top billiard player breaking a rack of billiard balls. His skill at this sets him in a good position to win the game but success is by no means a foregone conclusion.

The brief and incredibly rapid expansion during the inflationary epoch moved the space of the universe from dimensions measured in fractions of a micron to the grand size of a grapefruit. It was a single event that fixed for all time the amounts of matter and antimatter in the universe, and it set this matter in motion at a time when everything was sufficiently confined to be able to influence the future flow of matter and energy. This influence established the structure of the universe that we measure today. The beginning of the universe marked the start of space, matter, and time. The basis for all the material of the universe that will ever exist was formed at the beginning. As the universe expanded, the ethereal substance created at the beginning congealed into nuclear particles such as neutrons and protons and finally into the tangible matter we know today.

How does ancient biblical scholarship comment on this early period of the universe? At the start, the universe was concentrated into a speck of space smaller than a grain of mustard. That was the totality of all existence. Space and time both had their beginnings at this instant. A one-time phenomenon occurred that started the very young universe on its life-directed course. Biblically, this phenomenon is referred to as "the wind of God" (Gen. 1:2). In the ensuing expansion of the universe, the ethereal, nonsubstantive stuff of the creation took on the form of matter as we know it. The creation produced all that was, is, or will ever be extant in the universe. Even an avowed secularist must admit that this traditional view is a very close match of the view held by current cosmology.

Although photons (the stuff of which light is composed) were (and are) the main components of the universe, the young universe was dark. The photons were held in a confused soup of random collision with the masses of free electrons. Only after some hundreds of thousands of years did cosmic temperatures fall to a level that permitted electrons to bind in orbits about atomic nuclei. With this binding of electrons into atomic orbits, the ubiquitous photon-electron collisions ceased. Photons were free to travel. They burst forth bathing the universe in a blaze of light.

The Bible reveals a similar account. At the creation, we are told, the universe was dark, although it was packed with energy. The traditional description of this energy is that it was a "black fire." Light and matter were mixed together for an undefined time. Only after the expansion of the universe had proceeded, did light separate from the dark (Gen. 1:3).

In the early universe, the chaos of exquisitely hot and, therefore, turbulently mixed matter was matched by the highly ordered state of the universe's energy. The order of the energy was a result of its being concentrated into the small volume that was the universe at that time. In certain locations of the universe, as the universe expanded, there was a flow from chaos to cosmos. The cooling, which characterizes all expansions, let matter organize first into atoms and then into the massive systems we refer to as stars and galaxies. Through the alchemy of nuclear furnaces within stellar cores, matter was processed into the elements needed for life. Over the eons, on Earth and perhaps on other planets circling other stars, the symphony of life arose. It was a majestic order within what once had been a random mix of nuclear and atomic particles.

Biblically, we also see a flow from chaos to cosmos. Our study of the evening and morning sequences in the first chapter of Genesis has taught us that days and nights are not the main topic of conversation here. Each "day" marked an epoch, a flow from disorder toward increasing order in the material of the universe. This transition from disorder toward order is hinted at in the evening to morning phrasing of the biblical text. The root meaning of the Hebrew word for evening is "disorder" and for morning is "order." "And there was evening and there was morning" is telling us that in each "daily" episode, at a specific location within the universe, order was imposed by God on the disorder that had existed.

In a brilliant insight into the quality of the world present at the close of the six days of Genesis, Onkelos translates the "and it was very good" of Genesis 1:31 as "and it was a unified order."

The physical universe was prepared and mankind was in place. The social evolution that was to follow was no longer dominated by the inherent forces of matter that until now had played a major role in the world's development. From this time forward, there would be an interplay between man's free will and his knowledge of the will of the Creator. This interplay would be imprinted on future events.

The findings of cosmology date the creation at some 15 billion years ago. Biblically the age of the universe is the six days of genesis prior to Adam plus the 57 centuries since Adam. But tradition has also taught us that the Bible talks in the language of mankind. Can mankind comprehend billions of years? Not likely today and even less likely at the time of Moses. It took Einstein and the law of relativity to teach us that there is no absolute passage of time. It is as flexible as the possible differences in the force of gravity and the speed of motion across a boundary separating the observer from the observed. Within the fledgling universe prior to the formation of the Earth, there were innumerable such boundaries separating the multitude of stars that would eventually contribute parts of their mass to the Earth. The remnants of a myriad of supernovas may have supplied matter to the nebula that was to become our solar system. Because each of these stars had its own particular mass and velocity, each had its own cosmic clock that ticked at a different rate from all the other stellar clocks of the universe. The passage of time on any one star could be as different from the passage of time on other stars as six days is different from 15 billion years. As such, there is no one correct age for the Earth or the matter contained therein. The duration of days or years or even billions of years is only a relative observation. It is only locally correct. Until the observer and the observed are joined in a single space-time frame, there is no one-to-one correlation. For the Creator and the created, the union of frames of reference occurred when mankind, represented by Adam and Eve, absorbed the image of God, some 5700 years ago.

Among the infinite variety of environmental conditions possible in a universe and on a planet, the physical laws of the universe and the situation of the Earth with its temperate climate, moderate volcanism, and abundance of liquid water and carbon are a nearly perfect match for the needs of life. And, as if on cue, life appears almost immediately on the newly solidified Earth. In fact, it appears so rapidly that, statistically, life's formation from inanimate matter of the Earth cannot be attributed to random chemical reactions. The organisms of life are simply too complex to be accounted for by chance occurrences in a bath of inorganic chemicals. The abrupt appearance of the first forms of life is a presage of the similarly abrupt changes in the forms of life that would follow. The staccato nature of the fossil record, punctuated with species extinctions and subsequent radiations of new species, leads as if it were being led to the eventual appearance of hominids and then of humans. It presents a saga that defies description as a smooth, continuous flow of survival of the fit.

Paleontologists, archaeologists, and mathematicians alike find no adequate theory based on phenomena known in nature to account for the immediate appearance of life on Earth. The erratic patterns constantly found in the fossil record are explained by catastrophic events, some of which originated in outer space. Biblical tradition expects an erratic trend. As the will of the Creator is imposed from time to time on the otherwise natural flow of events, abrupt changes are introduced. Biblically, Divine punctuation in the development of the world is most evident from the tenfold repetition "and God said . . ." in the 31 verses of the first chapter of Genesis.

With this brief review of our physical and theological genesis in mind, we can answer some of the questions we posed. For the secularist, the events of the universe must by necessity form a cohesive set. This is simply because if we take the entire universe as the system under study, there is nothing else. The events of the universe are inescapably bound by a common cause. The laws of nature, subject to the fundamental nuclear, electromagnetic, and gravitational forces, are all that exist to have guided the transition of matter from its chaotic, ethereal past to the highly organized phenomena of life.

Ironically, except for the beginning of the universe and for the fact of life, the laws of nature might be sufficient to explain all that we see in the universe. But the laws of nature cannot account for the initial creation, nor can they account for life's almost immediate appearance on Earth after the Earth's formation. Only if we accept the most highly improbable sequence of random events as adequately possible can chance have produced life. But to do this is to make a mockery of statistical analysis. Except at the nuclear level, where quantum mechanics can alter statistical probability, the very laws of physics that predict the formation of stars, galaxies, and elements rely on the occurrence of the probable over the improbable. Without this there is no basis for physics. It is as if we were to say that a book on a shelf will of its own accord jump upward. This, in fact, *can* occur and if we wait long enough it *will* occur. We need only the patience to wait for the random vibrations of all the molecules from which the paper is composed to align in the upward direction. This might occur once in a myriad of billions of years and therefore that rare occurrence might occur right now, but it won't. Although *in the-ory* an event may occur, statistics have told us that *in reality* when the probability of an event occurring is very, very small, then there is essentially zero chance of it occurring.

Because the laws of nature cannot account for the appearance of life on Earth, we are left with an arbitrary—not cohesive—set of events in the flow that led to life. But there is *no place* for an arbitrary set here. The universe, when considered in its totality, is a single entity. The events *must* be related.

The arbitrariness appears only because we have used an un-guided, or secular, physics as the causal explanation. The real-ization that the events of the universe must belong to a cohesive set has forced many scientists to seek extraterrestrial inputs to account for life on Earth. The Nobel laureate Francis Crick has suggested that life is so complex that its seeding on Earth may have been essential for life's start on Earth. The eminent scientist Sir Fred Hoyle asserted that this seeding *must* have occurred and it was by a Divine force or superforce. A divinely guided physics provides the needed, single causal explanation of the flow of the universe.

Are we so complex, as our second question asks, that our ex-istence and the existence of our uniquely well-suited universe is so improbable as to require a causal explanation? I do not believe that we can use our complexity or the wonders of the universe as a basis for proof of Divine guidance in our formation. If you observe an apple blossom and attempt to predict the exact apple that will grow from that blossom—its shape, weight, and sugar content—there will be approximately zero possibility of your succeeding in this prediction. But an apple grows and it has a definite shape, weight, and sugar content. Its appearance is not extraordinary, other than the fact that life itself is extraordinary.

A myriad of forces combined to form this particular and unique apple just as they do to form a human or did to form the universe. It is naive to say in retrospect that the odds against producing the particular apple were enormous and therefore God must have shaped it exactly as it is. An apple was to form from the blossom, and it formed. How it did could be pure DNA direction or Divine intervention. Population genetics and molecular biology have taught us that a near infinite number of possibilities are held in the genetic material of each blossom. Neither in the size of the apple nor in the uniqueness of the universe is there *proof* of a Divine force. One might merely argue that if our universe were less marvelous, we simply would not be here to ask these questions.

Because the traditional biblical descriptions of the events of the early universe and young Earth have come down to us only in brief form, it is not possible to "prove" that these theistic traditions actually foretold the details of the recently rediscovered facts of cosmology. As usual, accepting or rejecting belief in the Divine must be a personal choice.

It is, however, remarkable that the theistic description, which was developed millennia before, and therefore in isolation from, the current scientific description, so closely matches the broad perception of current cosmology and paleontology. It is not that theology responded to modern scientific discoveries. Theology presents a fixed view of the universe. Science, through its progressively improved understanding of the world, has come to agree with theology.

For many people, and perhaps for most amid the sophistication of the Western world, the unhindered, objective laws of physics describe the universe with adequate precision. The layperson assumes that, although he may not understand the flow of nature from its pristine beginning to the world of today, there are scientists who have solved this riddle. We have become accustomed to relying unquestioningly on the ability of experts to supply the answers. No need to worry about tubercular cows as we pour milk over our high-fiber breakfast cereal. The scientists provided the biology and the engineers devised the technology for getting wholesome milk onto our table each morning. The problems are solved. We may not understand how they are solved, but we can sleep easy knowing that there are specialists out there somewhere who do know.

And so it is with the Big Bang and the evolution of life and a host of other problems related to our existence. We may not know *how* the Big Bang blasted us into being or how the fossil record demonstrates the development of life, but we can rest easy. The specialists know these answers.

Unfortunately, as we have learned in our journey through this book, at the key junctures the specialists do *not* have the answers. The fact is that the laws of nature as we experience them *fail to* account for the forces that in effect formed the universe. These same laws, which form the basis of our lives, also *fail* to account for the most central phenomenon of our lives, the appearance of life itself.

To the objective observer, this demands that a search be made elsewhere to find the causes that underlie our existence. Biblical tradition and current cosmology, two systems of thought that have *developed independently*, provide similar answers to several complex questions. This similarity warrants a causal explanation. A Divine origin for the biblical tradition would embrace the laws of physics and explain this similarity. A secular origin of this same tradition cannot account for tradition's prescience of the conditions of the early universe so recently rediscovered by science.

The demonstrated fact that the Bible contains explicitly apparent and complexly hidden information in ways that have been found in no other text places the Bible in the realm of the mystic. The uniqueness and accuracy of this multifaceted information may just provide the platform for fixing it in the Divine.

How you view this tradition is your choice.

CHAPTER 11

An Omnipotent God in a Limited Universe

■■■■■■■■■■■■■■■■■■■■■■■

Einstein once asked the question: "How much choice did God have in constructing the universe?" [Hawking:] "If the no-boundary proposal is correct, He [God] had no freedom at all to choose the initial conditions."

—A Brief History of Time

God: "Let me alone that my anger may wax upon them and I will consume them. . . ." Moses: "Remember Abraham, Isaac and Israel to whom you promised to multiply . . . and to give this land." And God relented of the evil which He said He would do to His people.

—Exodus 32:10–14

When we study the history of the universe in the light of the discoveries of science, we are confronted with several uncomfortable realities. There is the mathematically complex singularity of infinitely high material density and infinitely small physical dimensions at the instant of creation (or if you prefer, at the instant in which the current expansion of the universe started). Then there is the extraordinary improbability of life. Not only is life itself improbable, but its appearance almost immediately following the solidification and cooling of our once-molten planet defies explanation by conventional physical laws. The punctuated fossil record is no less easy to explain. The fossil record is now 100 times more elaborate than during the lifetimes of Darwin and Wallace. The increase in the number of known fossils has revealed even more jumps in the flow of life from the simple to the complex than were known in

Darwin's time. The very basis of Darwin's theory of evolution is that nature does not make jumps; yet the jumps are most prominently there for all to see. Punctuated equilibrium is now the accepted theory for many paleontologists. Smooth evolution is admittedly inadequate as a theory.

Do these circumstances mean that we must acknowledge a need for a Creator if we are to explain our existence? Or is our existence —and that of a universe suitable for life—an inevitable phenomenon? The answer to these questions is so subjective that, in 1951, Pope Pius XII said, "True science to an ever-increasing degree discovers God as though God were waiting behind each closed door opened by science." In 1954, George Wald assessed the findings of science, at least those within the discipline of the biochemistry of life, in almost exactly the opposite light: "Time is in fact the hero of the plot [in the generation of the first form of life]. . . . What we regard as impossible on the basis of human experience is meaningless here. Given so much time, the 'impossible' becomes the possible, the possible probable, and the probable virtually certain. One has only to wait: time itself performs the miracles."[1]

Let us assume that, at the least, there is potential room for a Creator to fit within the workings of the universe as described by theories of current cosmology. What would be the characteristics of such a force? Would God be so limited by the natural laws of the universe that this Creator would no longer be omnipotent, and therefore no longer compatible with traditional concepts of God?

Life as we know it is so special, so complexly organized and so fragile that it can flourish only within the narrowest environmental conditions. Carbon and oxygen and hydrogen are required in abundance to form the intricate and varied molecules of life. Yet neither carbon nor oxygen is abundant in the universe, nor were they produced in the Big Bang. The universe that was to nurture life needed the alchemy that would change the primeval building blocks of hydrogen and helium into the heavier elements of life. Water in its liquid state is needed as the medium within which the reactions of life were to occur. Yet liquid water occurs only within a narrow 100°C range. A long-term source of energy with constant output for billions of years is necessary to warm the water and to fuel the development of life from the simple to the complex. Such a long-term energy source can almost only originate with a star, but accompanying the warming stellar light is the devastating flux of ultraviolet and cosmic radiation. The potential home of life required a window that allows light to enter but keeps out the uv radiation. This same home must have an umbrella that effectively deflects the continual shower of lethal cosmic radiation.

These constraints deal only with the macroscopic characteristics of the universe. At the nuclear level, the demands of life are equally precise. The forces that bind protons and neutrons into nuclei of atoms must be sufficiently strong to form the stable units we refer to as the elements, but weak enough to allow the spontaneous fission of some of the nuclei of these elements. This fission fueled the volcanoes that released trapped vapors and gases and in this way formed the biosphere, the thin film of water and air in which we and all life thrive. Electromagnetism, which binds electrons and thus defines the chemical properties of the atoms, is also finely balanced. It must be weak enough to free electrons for passage to neighboring atoms but strong enough to organize and join these adjacent atoms into ions and molecules, the basis for the solid and liquid structures of all matter. Gravity, the most enigmatic of the four fundamental forces of the universe, is the weakest of the four, yet at large distances it is the most powerful force. As such it shapes the macrostructure of the universe. If gravity were significantly more powerful, the life of the stars would be too short to allow biological life to flourish. Increasing or decreasing gravity, which joins planets and stars and galaxies in their flight through space, would result in unstable orbits, with planets spiraling toward, or away from, their star.

The blend of the natural laws of the universe, along with the magnitude of the fundamental forces that operate among matter and energy, cannot vary by much if the universe is to develop such complicated things as multibillion-year-old stars and living cells. Are God's hands tied by this narrow, yet natural, framework within which life must fit?

In a sense, the answer is yes. To get to man, through the temporal development of a universe and a biology, there is not much opportunity for variation in the physics of the universe. Of course, this God might have chosen not to create a universe capable of bearing mankind. Then there would be no questions asked about such constraints on God. We would not be here to ask them!

It might also be argued that, not only is the type of universe that can bear mankind limited in its options for design, but also our very existence is *essential* for God. Although the very basis of the biblical Creator is that of an omnipotent force, an omnipotence having no external needs, there are characteristics of the Creator, as evidenced in the universe, that make it appear that our existence is the result of a godly desire.

When Moses sought to understand God, he beseeched God to "inform me of Your way [literally, *path*] that I may know you" (Exod. 33:13). The qualities of an omnipotent and infinite God are unknowable and so to fulfill the request, God told Moses, "I will pass *all My goodness* before you" (Exod. 33:19).

The meaning of this passage is difficult to discern. What is this "goodness"? Certainly it cannot refer to a physical image of God; God has no body. The implications contained within the word *goodness* may be learned from Genesis 1:31 where *good* is also used in relation to God. Literally this verse reads: "And God saw everything that He had made and behold it was very *good*." Here *good* refers to the quality of God's creation: the universe, life, and mankind.

From the parallel use of *good* in Exodus 33:19 and Genesis 1:31, it appears that "all My goodness" refers not to a quality of God but rather to the very nature of the creation, the exquisitely balanced, or orderly, interrelationships therein, as Onkelos translates Genesis 1:31: "And God saw everything that He had made and behold it was a unified *order*."

"I will pass all My goodness [read here, all My creation] before you." The very act of creation and the existence that flows from this creation is good.

Because the circumstance of existence is divinely considered to be good, then if a universe, regardless of its form, represents the primary form of material existence, God was in a sense *bound* to create a universe that could maintain a wide variety of existences. The form of our universe satisfies this Divine criterion, and does so in what may be the most expansive way possible. The entirety of the Bible and the entirety of cosmology describe a genesis of existence, a development, from the amorphous void of Genesis 1:2, through to the multifaceted community of today's universe. It is in this meaning of confinement that God may have been confined to create not only *a* universe, but to create *our* universe!

Have we not confronted here a contradiction in terms? The very concept of a Creator being constrained to perform an act is an infringement on the inherent omnipotence of the Creator. The Torah did not shy away from confronting such paradoxical situations. When a monarchy was to be established in Israel, the Bible clearly defined the rules and limitations that governed the king (Deut. 17). What could be more absurd than a king with limitations! The legal omnipotence of the regal was taken as an obvious and inherent trait of that position. England was to wait 2700 years to accept this paradox as nonparadoxical. It was not until June 15, 1215, at Runnymede that King John placed his seal on the Magna Carta and so limited the powers he and all future kings in England would hold over their subjects.

It may come as a surprise that the Torah also describes limitations, not only of monarchs, but also of God. These limitations are, however, there. They are not inherent characteristics of the Creator. They are traits assumed in God's manifestation in the universe. It is these traits that we are to emulate if society is to achieve its maximum potential.

A sufficiently sensitive and patient person might closely study the varied aspects of the universe and accomplish this goal of understanding God's way. The sages say that the Patriarchs, Abraham, Isaac, and Jacob, fulfilled the laws of the Torah. Yet Abraham is described as serving milk and meat to his guests in the same meal (Gen. 18:8), an act forbidden by rabbis living in later times. Some biblical interpreters, disturbed by this apparent breech of law, go to great lengths in attempting to explain how this could possibly have been done. Similarly, Jacob married Leah and Rachel, two sisters. This is explicitly forbidden in the Bible (Lev. 18:18).

When the sages said that the Patriarchs fulfilled the Torah, the implication is not necessarily that these individuals and their disciples carried out each detail of worship or law. The Torah that the Patriarchs observed embraced the fruits that spring from observing the individual laws, fruits that for most people are accessible only through the guiding laws of the Bible. This is in keeping with the root meaning of the word *Torah*, which is "instruction, showing, or teaching the way." *Torah* does not mean "law"; this is a common mistranslation.

The laws of the Torah are an aid, a guide to the path that leads toward the much larger goal of instilling harmony among mankind and between the creation and the Creator. The great principle of the Torah is to love your neighbor as yourself. The Patriarchs, in their sensitivity to their surroundings, were able to understand these goals and to honor them, even though they did not possess the actual text of the Torah.

Unfortunately, most of us are not endowed with sensitivity and patience in amounts commensurate with the task of deciphering from the ways of the universe the path of God. For us, the Bible provides the needed, prepackaged descriptions. It is from these that we must take our clues for life's guidance.

The laws of the Torah, especially those dealing with the relations among mankind, are in effect a description of the acts of the Creator toward the creation. In Deuteronomy 6:18, they are summarized as doing the "just and the good." Why the "just" and the "good"? Isn't justice sufficient? Apparently the author of the Bible did not think so. Justice alone is not enough. It must be tempered with goodness. This is the essence of the attributes of God revealed to Moses as God's goodness passed before Moses. "The Lord, the Lord, mighty, merciful and gracious, long-suffering and abundant in kindness and truth, keeping kindness to the thousands, forgiving iniquity, transgression and sin, and clears [the repentant] but who will not clear [the unrepentant]" (Exod. 34:6–7).

Because we can read the ways of God in the ways of the universe and also in the principles of the Bible, then God's management of the universe must be consistent with the principles of the Bible. Divine tenacity to these postulates is, of course, the very essence of the biblical God. It was the concept of fickleness in pagan deities that drove the disciples of these deities to extremes of bodily mutilation (1 Kings 18:28) and child sacrifice (Deut. 12:31). The barbaric acts were no more than frantic attempts to find means of satisfying what was perceived as ever-changing whims of their gods.

God's adherence to the norms of the Bible in effect limits Him; He is bound to do the "just and the good." Abraham used God's fidelity to successfully argue with Him against the destruction of the righteous people of Sodom (Gen. 18:23–32). Moses also relied on the fixed way of the Creator. When confronted by the people's Golden Calf, God threatened to destroy the masses that Moses had led from Egypt (Exod. 32:7–14). Moses, having the Torah in his possession, reminded God of the Divine promise to make the progeny of Abraham, Isaac, and Israel into a great nation and to give them the land of Canaan and its surroundings. God certainly had the power to destroy these people, but that would be contrary to the Divine promise. If the promise to the ever-faithful Abraham was to be honored, and this is inherent in the relationship between God and Abraham, then the children of Israel might be punished but not annihilated (Gen. 32:33). God relented in accord with Moses's request.

Once more, during their trek through the desert, the Israelites rebelled. Again they are threatened with Divine destruction. Moses reasons with God, "Now if You will kill this people as one man, the nations that have heard Your reputation will say because God could not bring this people to the Land which He promised to them, therefore He slayed them in the desert" (Num. 14:15–16).

Presented with the specter of the surrounding nations claiming that Israel's destruction resulted from God's inability to bring them into Canaan, God relented (Num. 14:20–24). Again, because of a Divine intent, in this case the manifestation of God's power within the universe, the Divine options were limited.

Because we are only able to view the magnitude and breadth of God's power in the context of our world, the infinite extent of this power is not within human comprehension. Moses was God's intimate servant and even he failed to perceive its extent. "And the children of Israel also wept and said, who will give us meat to eat? We remember the fish we ate in Egypt for nothing, the cucumbers and the melons, the leeks and the onions and the garlic" (Num. 11:4–5).

It was no longer relevant that, while they ate these delicacies, they "sighed from the enslavement and cried and their cry rose to God by reason of the bondage" (Exod. 2:23). What mattered now was that the manna was not meat.

And God said to Moses, "Say to the people sanctify yourselves for tomorrow and you shall eat meat for you have wept in God's ears saying who shall give us meat for it was good for us in Egypt. God will give you meat and you shall eat. Not for one day shall you eat, and not for two days, and not for five days, and not for ten days. But for a month of days...." And Moses said, "The people within whose midst I am are six hundred thousand footmen and You said I will give them meat that they may eat for a month! If flocks and herds are slaughtered will it suffice them? and if You gather for them all the fish of the sea shall it suffice them?" And God said to Moses, "Is the hand of God short?"

(Num. 11:18–23)

The simplicity of God's reply tells of the sublime nature of the Creator's power. Is the hand of God short? Is there anything beyond the reach of this power? The answer is no. Moses could not comprehend this because mankind cannot quantify the infinite, either mathematically or philosophically.

And so it is not surprising that scientists might find within the functioning of the universe aspects that appear as limitations to God's power. This stems, not from an attribute of an infinite God, but rather from the attributes of man; namely, that we are able to discern characteristics of the Creator only as they are related to that which is created. This limitation of ours immediately reduces what might be infinite to the finiteness of our existence. Theologically, there is no problem in accepting constraints to Divine options if the Divine wants to run the universe according to stated and established laws.

We are confronted with a further problem. Having established principles for the operation of the universe, was the universe "wound up" and left to run its course undisturbed from outside interference? Two of the scions of the Darwinian theory of evolution believe they have found proof that indeed the universe runs a course bound only by the laws of nature. Ernst Mayr and George Gaylord Simpson, men of impeccable academic credentials, base their proof on the evidence of paleontology.[2] They claim that examination of the details in the fossil record of any evolutionary process, such as the trend toward a specific skeletal feature, reveals a flow that is not at all smooth or unidirectional. Changes in trend, reversals, and even extinctions in the trend are so common in the fossil record as to be the rule rather than the exception. Were there a Divine teleological guide to evolution, then reversals would imply errors or changes in God's initial planning. To use the terminology of Howard Van Till, this would indicate that the creation was functionally incomplete. This, Mayr and Simpson claim, is certainly inconsistent with the traditional concept of an omnipotent Creator.

(Why there should be evolutionary development at all is not a theological problem. The development of the world through time is taken as a given. This is the record of the opening chapters of Genesis.)

But have Mayr and Simpson really discovered an inconsistency that runs counter to biblical tradition? "And it came to pass when Pharaoh had sent out the people [of Israel] that God did not lead them through the land of the Philistines, though it was close, for God said lest the people regret when they see war and return to Egypt. And God turned the people toward the way of the desert of the Sea of Reeds" (Exod. 13:17–18). This alteration in course misled Pharaoh. "Pharaoh will say concerning the children of Israel, they are confused by the land, the desert has closed them in" (Exod. 14:3).

It also misled the Israelites. "And when Pharaoh drew near, the children of Israel lifted up their eyes and behold Egypt travelled after them and the Israelites feared greatly and shouted to God. And they said to Moses, 'Because there were no graves in Egypt you took us to die in the desert' " (Exod. 14:10–11).

But both the Egyptians and the Israelites had misread the purpose of the circuitous course that was being followed by the just-liberated slaves. The journey was not as haphazard as perceived. "And Moses said to the people, 'Fear not, stand and see the salvation of God which [God] will do for you today' " (Exod. 14:13). As it turns out, the people had to believe in this potential for salvation and make the first move if they were to have this salvation. "And God said to Moses, 'Why do you cry to me? Speak to the children and go forward. You raise your rod and stretch your hand over the sea' " (Exod. 14:15–16).

There were multiple goals in the exodus from Egypt, not all of which were obvious. A portion of these goals could best be fulfilled by directing the journey along a path that retraced its steps. "Pharaoh will say concerning the children of Israel, they are confused by the land, the desert has closed them in . . . . That he [Pharaoh] shall follow them, and I [God] will gain honor by Pharaoh and all his army that Egypt will know that I am the Lord" (Exod. 14:3–4). This goal, stated clearly in Exodus 14:4, was obvious neither to Pharaoh nor to the Israelites. But it was part of the Divine strategy from the very start of the exodus. It was not an addendum stapled on to correct a plan that was initially functionally incomplete.

Because most of us cannot discern the Divine in nature, we depend on the Bible to understand the ways of the Creator. Neither limiting the options for the perceived actions of God in the formation of a universe nor the revelation of a circuitous paleontological record is in contradiction with the traditional concept of God's role in the universe. If scientists want to find proof for the absence of God in the universe, they will have to seek more profound evidence.

It is of interest that classic philosophers of the caliber of Pierre-Simon de Laplace (1749–1827) also believed that the laws of the universe confine God to the few options able to produce the complexity of life. Because of these constraints, Laplace argued that by knowing the conditions of the universe at any one time and knowing the laws that govern the functioning of the universe, one can predict all future events. Free will was thus eliminated. There was and is only one predetermined path that can lead from the past to the present and onward to the future. Because every act or event is a reaction to preceding ones, if we could but measure exactly the state of the universe today, we could apply the laws of nature and know the future. This is the philosophy of determinism. In theory at least, it seemed to be obviously true, even though it infringed on the role of God in the universe. It was not until the early 1900s, 150 years after Laplace formulated his theory, that the uncertainty principle stated a limit to the precision by which the condition of any particular particle can be measured. With the realization that uncertainty was an inherent aspect of the universe, determinism crumbled. Because the condition of the universe could never, even in theory, be exactly defined, the future could never be determined.

Recently, the predictability of the future has been found to be even more limited than imagined by the uncertainty principle. The uncertainty principle deals with systems at the nuclear level and with uncertainties in positions and velocities commensurate with this microsize world. Perhaps at a large scale, say at the size of humans and larger, we might still be able to predict what the future will hold. Alas, this too is not possible.

With the best of instruments, minute uncertainties remain in the location and velocity of even macrosize objects. These uncertainties are amplified each time the object under study interacts with another object or force. In a very short time, this amplification enlarges the original uncertainty to a magnitude that equals or exceeds the size of the system being measured. Because the purpose of predictions is to describe events following sequences of interactions (we can easily predict the future position of an object at complete rest and not having interaction with the environment; it remains in place, but such information is not very interesting), the predictive process, due to the amplification of uncertainty, rapidly fails to predict reality. Future events in the real world are exquisitely sensitive to present conditions.[3]

What does this mean for planning your next picnic? Edward Lorenz of M.I.T. is one of the originators of this recent study into the amplification of uncertainty. He has shown that, in spite of satellites and a multitude of Earth-based stations all gathering meteorological data, predictions of weather beyond two or three weeks bear no more resemblance to reality than would a random choice taken from a pile of old weather maps.[4]

What the amplification of uncertainty means for philosophers, theistic or secular, is that our future is unknowable.

We live in a world in which existence in and of itself is good. The less than perfect condition of our existence is not an inherent aspect of the world. The entire flow of matter over the eons has been, in our corner of the universe at least, a journey from disorder toward harmony. So much is already good that the bad stands out by contrast. This gives cause for optimism. We may yet win the race against the idea that might makes right. This opportunistic philosophy, the idea that power brings with it the justification for dominance, is a sort of social Darwinism. But its reliance on biology for its legitimization is unjustified. Altruism, the phenomenon in which one animal helps another and receives no obvious benefit in return, is widespread in nature and even occurs between individuals of different species. Such giving behavior is the antithesis of opportunism.[5] The posture of exploitation and elitism instructs civilizations to travel the path that has in the past and will in the future lead only to intellectual decay and physical ruin. Curiously, opportunism extols exactly the opposite qualities that are most esteemed in Moses: "The man Moses was very humble, more so than all mankind on the face of the Earth" (Num. 12:3)—and in mankind in general—"It has been told of you, man, what is good and what God requires of you: only to do justice, love kindness, and to walk modestly with your God" (Mic. 6:8). In fact, the Talmud states unequivocally that "in each place that you find the greatness of God, there you find God's gentle kindness."[6]

As individuals within our human society, we can choose a more compassionate basis for our instruction than that provided by opportunism. In so doing, we can extend the already abundant goodness to all our surroundings. It is for this that we are possessors not only of a free will that allows us to move in accord with our individual desires. All animals have this capability. But beyond the freedom of will, we members of humanity are endowed with the sensitivity and the special knowledge that enables us to distinguish the true from the false and, at times, the good from the evil.

> Then the eyes of the blind shall be opened and the ears of the deaf unstopped. The lame man shall leap as a deer and the tongue of the simple sing. . . . The people that walked in darkness have seen a great light; they that dwelled in land of images, upon them has light shone
>
> (Isa. 35:5–6, 9:2).

Notes

1. Wald, "The origin of life," *Scientific American* 191 (August 1954): 48.
2. Mayr, in Dobzhansky et al., *Evolution*, p. 6.
3. Gleick, *Chaos: Making a New Science*.
4. Gibbons, "Chaos and the real world," *Technology Review* 91 (July 1988): 10–11.
5. For a survey of the occurrence of altruism in nature see the following: Gould, *The Panda's Thumb*; Taylor, *The Great Evolution Mystery*; Smith, *The Evolution of Behavior*; and Wilkinson, "Food sharing in vampire bats," *Scientific American* 262 (February 1990): 76.
6. *Babylonian Talmud*, Section Megillah 31a.

Epilogue

■■■■■■■■■■■■■■■■■■■■■■

With the help of ancient theology and modern cosmology we have completed our trek from the pristine beginnings of the universe to the appearance of mankind on Earth. Skeptics on both sides of the theological aisle may have raised their eyebrows at the information presented; we can become so accustomed to our personal dogma that we often fail to realize that the opposition exists because it too has a realistic basis for its beliefs.

For the Bible scholar, it is not an easy task to accept as reality that for the past 100,000 years there existed animals such as hominids and that the skeletons of these ancient animals are near replicas of those of modern man. But the fossil evidence is abundant and irrefutable. It is folly, no it is counterproductive, to close one's eyes to this fact.

The skeptic must find it equally difficult to acknowledge the overwhelming improbability of life arising from inert matter through unguided chemical reactions. But it is a tool of the scientist, the mathematics of statistical analysis, that has proven this improbability of life.

The fact is that the findings of science are a challenge to believer and secularist alike.

Similarly, biblical tradition argues with all who restrict themselves to a literal understanding of that which is contained in the Bible. Such an approach only deprives the skeptic, agnostic, and believer alike, of a wealth of information. The Bible talks in the language of man. By necessity. It had to be meaningful to the just-freed slaves standing at Sinai, while retaining the depths of meaning intended for generations yet to be born. Once we accept this as fact, then interpretations of the Bible that seek out subtleties in wording and form can be seen as far more than fanciful pipe dreams.

The biggest error of his life, Einstein said, was insisting that the universe was static and not expanding. Had he had the benefit of biblical tradition, he might have felt secure in the predictions of his theory of relativity, that indeed the universe was in a state of expansion. The 3.5°K cosmic radiation background confirms not only the scientific theory of the expanding universe. It also confirms the traditional biblical statement of the expanding universe.

The fossils of Cro-Magnon civilization stretch from France to the Ukraine. The argument that the Flood in the time of Noah altered the fossils and made them useless as archaeological tools is hardly credible to a generation of biblical scholars who have studied the physical sciences as well as the Bible. But such arguments are needed only by those unwilling to accept the findings of traditional biblical interpretation. The existence of pre-Adam animals with shapes and intellect similar to humans was discussed 1000 years ago by biblical sages just as it has been discussed during the past 100 years by archaeologists. The data are not a threat to either side. It was not by chance that the biblical calendar is dated from the appearance of Adam and not from the creation of the world. Relativity has proven the flexibility of time during those six, pre-Adam days of Genesis.

If we are ever to reach an understanding of our cosmic origins, an understanding that is compatible with *all* the information we have, then skeptics looking over each of my shoulders must look carefully into each other's texts as well.

Appendix

■■■■■■■■■■■■■■■■■■■■■■■

THE EFFECT OF MOTION ON THE TRAVEL TIME
OF AN EMITTED ENERGY WAVE (SEE CHAPTER 2)

Michelson and Morley attempted to measure how the Earth's motion through an imaginary ether affects the speed of light. Because the Earth orbits the Sun at approximately 30 kilometers a second, measuring the speed of light parallel to and perpendicular to the direction of the orbit should show a difference in the speed of light if light followed the same rules of travel that other waves follow.

To make this clear, let's take an example with which we are familiar. When sound is emitted in still air, the sound moves out from its source at a fixed speed regardless of the velocity of the source relative to the air. As an example, for standard sea-level conditions, when the horn is blown in a parked car the sound of the horn travels from the car at a speed of about 300 meters per second. If the same horn is blown when the car is moving at 30 meters per second, the velocity of the sound remains 300 meters per second and not 330 (300 + 30) meters per second. The 300 meters per second is the fixed speed of transmission for sound energy in this air, just as the speed of transmission for electromagnetic radiation (such as light) is constant at approximately 300,000,000 meters per second. Now if 300 meters distant there is a smooth, rigid wall, the sound will be reflected by this wall. This reflected sound may travel back to the original source of the sound. That is what echos are all about. If the air is still, it will take 2 seconds for the echo to reach the original source (300 meters ÷ 300 meters per second to the wall + 300 meters ÷ 300 meters per second echo return from the wall = 1 + 1 = 2 seconds). And that is what the Michelson-Morley experiment showed for the light. The travel time of the light was always the same. But there was a crucial difference. In their experiment, there was supposed to have been an effect of the Earth speeding through the ether as the Earth orbits the Sun.

Let's see what the effect of wind is on the travel time of sound. Assume that a wind of 100 meters per second is blowing directly into the face of the sound emitter. Now the travel time for the echo to reach us will be 2.25 seconds [300 meters ÷ (300 − 100) meters per second to the wall + 300 meters ÷ (300 + 100) meters per second to return from the wall = 1.5 + 0.75 = 2.25 seconds]. The presence of the wind changed the round-trip travel time. Pilots who fly fixed routes are well aware of this effect of wind on their round-trip flight times. The motion of the Earth through the ether should have had a similar effect on the round-trip time for the light to travel in the Michelson-Morley experiment. There was no such effect.

For Einstein, this meant that there was no ether!

TRANSFORMATIONS OF MASS, LENGTH, AND TIME AS DESCRIBED IN THE SPECIAL THEORY OF RELATIVITY (SEE CHAPTER 2)

Five years after Michelson and Morley published their results, which demonstrated that the speed of light was not affected by the direction of motion of the light, George F. Fitzgerald suggested a rather unconventional solution that would explain their paradoxical results. Fitzgerald postulated that the ether through which the Earth travels, although unable to be perceived, has its effect on us. The pressure of this ether wind compresses matter along the direction of motion just enough so that the distance traveled by the light in this direction is a bit less than the distance traveled perpendicular to the direction of motion. The contraction, of course, cannot be measured. After all, our yardstick also feels the pressure of the ether and so contracts by the same fraction. The extent of contraction is proportional to the velocity of the object through the ether. In 1893, this fractional contraction was quantified mathematically by Hendrik A. Lorentz in what has come to be known as the Lorentz transformation:

$$X = X_0 \left[1 - \left(\frac{v}{c} \right)^2 \right]^{0.5}$$

In words, the transformation reads: The new length of an object (X) equals the original length (X_0) multiplied by the square root of 1 minus the object's velocity (v) squared divided by the speed of light (c) squared. Extraordinary! If v reaches the speed of light, then we have $1 - 1$. The dimension shrinks to zero!

Einstein, seeing the basic relevance of this form of transformation, developed the equations for the relativistic changes in mass and space dimensions.

The mathematical statements that describe the relativistic changes in mass, space, and time are all similar to the Lorentz transformation. Relativistic mass (m) equals

$$m = m_0 \left[1 - \left(\frac{v}{c} \right)^2 \right]^{0.5}$$

Relativistic length (x) equals

$$x = x_0 \left[1 - \left(\frac{v}{c} \right)^2 \right]^{0.5}$$

Relativistic time passage (Δt) equals

$$\Delta t = \Delta t_0 \left[1 - \left(\frac{v}{c} \right)^2 \right]^{0.5}$$

Relativistic speed of light (c) equals

$$c = c_0$$

That is, the speed of light remains constant regardless of the velocity (v). The energy contained as the mass of an object is represented by the familiar equation

$$E = mc^2$$

MASS-ENERGY RELATIONS: AN EXPLANATION OF
$E = mc^2$ (SEE CHAPTER 2)

The energy of the stars and the heat from nuclear decay of heavy elements within our Earth derive from the same phenomenon: the conversion of mass into energy. In nuclear fusion reactions, when as in the core of a star two hydrogen nuclei fuse with neutrons to form helium, the resulting weight of helium is slightly less than the sum of the masses of nuclear particles that joined to form the helium. The same is true of fission of heavy elements. When a uranium atom breaks apart and forms several lighter elements, the sum of the masses of these newly formed particles is less than the mass of the parent uranium. For both fusion and fission, the weight or mass lost during the reaction appears as energy and it appears in copious quantities.

The density of energy when in the form of matter is so great that converting a single gram of matter (any matter) into pure energy releases enough heat to boil 34 billion grams (approximately 30 million quarts) of water into steam. This equivalence between mass and energy is familiar to us as the extraordinary force released in nuclear weaponry. A 20-megaton hydrogen bomb, we pray exploded as an exercise in the peaceful uses of nuclear energy in space, gets its power (equivalent to 20 billion kilograms of TNT exploded in a single blast) by changing a mere 0.9 kilogram (approximately 2 pounds) of hydrogen mass into pure, exquisitely hot, energy. It is the identical reaction of the H-bomb, the fusing of hydrogen into helium, that powers our Sun and most other stars.

The energy released in exothermic chemical reactions also results from changing part of the mass of the reactants into energy. However, the density of the energy release is so small that weight balances are not sufficiently sensitive to detect changes in mass from before to after the chemical reaction. Even the very energetic chemical reaction of burning hydrogen gas with oxygen to form water vapor (we use this reaction to power rocket engines) does not convert mass into energy rapidly enough for us to measure the change directly. For each 18 grams of water vapor formed, only 2.7 billionths of a gram (or 2.7 nanograms) of the reactants (hydrogen and oxygen) are converted into their energy form. This is released as 57,780 calories of heat.

Bombarding the naturally occurring nitrogen isotope, nitrogen-14, with deuterons produces another nitrogen isotope, nitrogen-15, in an energetic reaction releasing 8.6 million electron volts of energy per atom of nitrogen-15 formed. For each 15 grams of nitrogen-15 produced, 9.2 thousandths of a gram (or 9.2 milligrams) of reactants are converted to their energy form. As is typical for a given weight of reactants, the energy released (and correspondingly the mass that converts to energy) in the nuclear reaction is about 1 million times greater than in the chemical reaction.

The million-to-one ratio in energy release between nuclear and chemical reactions has its basis in quantum mechanics. Chemical reactions are involved with changes in the arrangements of electrons orbiting a nucleus. Nuclear reactions deal with changes in the structure of the nucleus of atoms. The energy by which electrons are bound in a particular orbit is in the order of one-quarter of an electron volt while rearranging the particles of a nucleus involves energy changes in the order of 0.1 to several million electron volts. Hence the million-to-one ratio.

In the early moments of our universe, both the conversion of energy into mass and mass into energy were equally dominant. Today the results of mass converting into energy are seen as the light and warmth of the Sun and stars. The conversion of energy into mass is seen in photographs taken in high-energy physics labs.

BIBLICAL INTERPRETATION (SEE CHAPTER 1)

In our study of Genesis and the Big Bang, it is essential to realize that there are meanings to the Bible that are not apparent from a casual reading. The search to find meanings within the Bible that expand on the literal text has been a part of Western theology since its inception. These interpretations are based on more than merely the whim of the interpreter. There is a history and a basis to each insight brought forward by the scholars on which we are relying. The sense of these insights is carried by subtle variations

TABLE 5. THE RELATIONSHIP BETWEEN WAVELENGTHS, FREQUENCIES, AND "BLACK BODY" TEMPERATURES OF PHOTONS (ELECTROMAGNETIC RADIATION)

Energy Wavelength (meters)	Energy Frequency (cycles per second)	Black Body Temperature (°K)	Common Name Given to the Photon
10^{-15}	10^{23}	$<5 \times 10^8$	
			gamma rays
10^{-10}	10^{18}	10^7	
			x rays
			visible light
10^{-5}	10^{13}	1000	infrared
			radar
			microwaves
1	10^8	< 0.01	television
			FM radio
			AM radio
10^5	10^3		

in the grammar, by unusual patterns of letters, and even by nuances in the calligraphy of the Bible, the actual physical form of the text.

I offer here only two examples of the thousands of subtleties found in the Bible. I chose these two examples, not because they reveal profound secrets, but because they can be easily verified by layperson and scholar alike.

The written text that Moses received is divided into five books: Genesis, Exodus, Leviticus, Numbers, and Deuteronomy. In Hebrew, these books are collectively referred to as the Torah (Torh, in Hebrew). Take the first time the Hebrew letter T appears in the Hebrew version of the Book of Genesis. Count out 49 letters from that T and record the next letter, that is, record the 50th letter. Repeat this three times. The result: Torh. Do it again in Exodus. The same result: Torh. Do it again in Leviticus. The result: gibberish! However, take the first letter of the four-letter explicit name of God, the word Jehovah when transliterated into English using consonants for the Hebrew name of God and the vowels of the Hebrew word for Lord. Count 7 letters. Repeat this three times. The result: Jehovah (or JHVH in Hebrew). Now to Numbers. The

■

word Torh appears at the characteristic 49-letter spacing, but backwards. That is, it faces the Jehovah of Leviticus. Repeat the process in Deuteronomy and get the same result: Torh facing Jehovah. Torah brackets God's name and in effect always faces it.

TORH
(Torah in English)

תּוֹרָה
(Torah in Hebrew)

T
בְּרֵאשִׁית בָּרָא אֱלֹקִים אֵת הַשָּׁמַיִם וְאֵת הָאָרֶץ
——— 49 letters ———→

O
וְהָאָרֶץ הָיְתָה תֹהוּ וָבֹהוּ וְחֹשֶׁךְ עַל פְּנֵי תְהוֹם וְרוּחַ אֱלֹקִים
——→ ←——

מְרַחֶפֶת עַל פְּנֵי הַמָּיִם וַיֹּאמֶר אֱלֹקִים יְהִי אוֹר וַיְהִי
——— 49 letters ———

R
אוֹר וַיַּרְא אֱלֹקִים אֶת הָאוֹר כִּי טוֹב וַיַּבְדֵּל אֱלֹקִים בֵּין
——— 49 letters ———→ ←———

H
הָאוֹר וּבֵין הַחֹשֶׁךְ וַיִּקְרָא קֱלֹהִים לָאוֹר יוֹם וְלַחֹשֶׁךְ
←———

קָרָא לַיְלָה וַיְהִי עֶרֶב וַיְהִי בֹקֶר יוֹם אֶחָד

Why 49 spaces and then the next letter to spell the word Torah? In Leviticus 23:15, we are commanded to count 49 days from Passover and then to celebrate the holiday of Weeks on the next (50th) day. The holiday of Weeks commemorates the giving of the Torah on Mt. Sinai.

Why the 7 spaces to spell the name of God? The number 7 occupies a special place in the Torah. The Sabbath is the 7th day of the week, the 1st day made holy in the Bible. Traditionally, the Sabbath is a sign acknowledging that God created the universe.

The shape of the written text, its actual calligraphy, is a still more subtle form of information transfer. Had there not been a substantial tradition that this was truly a valid form of carrying information from generation to generation, it is unlikely that such care would have been taken in maintaining the integrity of the written form of the text.

In Numbers 9:10, we are informed that in the event of specific extenuating circumstances, a person could observe the holiday of Passover on the 14th day of the *second* month of the year, rather than on the date when the Exodus actually occurred, the 14th day of the first month, the springtime month of Nissan: "And the Lord spoke to Moses saying: 'Speak to the children of Israel to say: any man if he is impure by a soul or on a distant journey, for you or for your future generation, he shall make the Passover to God. On the second month, on the fourteenth day . . .' " (Num. 9:9–11).

There is nothing unusual here except for a single extraneous addition. In these verses, there is a dot, a spot of ink, *always* written above the final letter of the Hebrew word *rehokah*. The literal meaning of *rehokah* is "distant, far away." There was no need to dot the word if its import is to be only the literal sense. Something more must be intended. The meaning of "impure by soul" is taken to refer to a person recently in contact with a corpse. Prior to celebrating Passover, the impure person must purify himself. But contact with a corpse is not mentioned here, although it is mentioned a few verses prior to this passage.

Here the dotting of the word for distant might imply that an imperfection of the soul and the distant journey were somehow related. Rashi makes the terse comment on this verse, "The dot on it [*rehokah*] teaches [that *rehokah* means] not necessarily far off, but even if he was just outside the doorway of the court [of the Temple] all the time for the [Passover] sacrifice." This seems to be a rather liberal interpretation of *far off*. Why should the few steps needed to cross the threshold and enter the Temple court-yard be considered "far off"?

The sages realized that participation in the holiday that marks the beginning of the journey from slavery to the revelation at Mt. Sinai required more than a physical presence. Every participant, even today, must feel as if he or she actually went forth from slavery. To arrive at that level of spiritual participation takes preparation. If a person was spiritually "far off" on the 14th of

Nissan, 30 extra days were given to make that journey and arrive at the Exodus.

ODDS: CALCULATING THE LIKELIHOOD OF AN EVENT HAPPENING BY CHANCE (SEE CHAPTER 7)

In this book, I have cited several instances in which persons have claimed that life or certain protein molecules may have formed purely by chance over geologically long periods of time. To better grasp how unlikely it is that a complex event will occur by chance, it is instructive to calculate the odds of several such events occurring by chance.

Consider our decimal system. There are 10 digits, 0 through 9. In a stack of 10 cards, each having a different digit written on it, the chance of selecting any given number, for example a 3, is one in 10. If we have 100 cards, with 0 through 99 written on them, the chance of selecting a 3 is one in 100. With 1000 cards, the chance is one in 1000. Note the relationship between the odds and the way the number changes. At each digit in a number, there are 10 variations (0 through 9). With a one-digit number, the odds of choosing a certain number are one in 10^1. (The exponent shows the number of zeros following the one. Hence 10^1 equals 10.) With a two-digit number, the odds are one in 10^2. With a three-digit number the odds are one in 10^3. The odds change with each added digit by a factor of 10 because we are in a 10-base system.

Our alphabet has 26 letters. That is, it is a 26-base system. Therefore with each added letter, the odds of choosing a specific letter decrease by a factor of 26. The chance of choosing a given letter, a B for example, from among all the letters of the alphabet is one in 26. The chance of randomly spelling a certain two-letter word is one in 26×26 (written in scientific notation, this is 26^2). That equals one in 676. With a three-letter word, the chance is one in 26^3 (one in 17,576).

Now let's examine the frequently mentioned scenario in which monkeys type random letters. In referring to the possibility of chance forming our universe and us, Stephen Hawking, in *A Brief History of Time*, writes, "It is a bit like the well-known hordes of monkeys hammering away on typewriters—most of what they write will be garbage, but very occasionally by pure chance they will type out one of Shakespeare's sonnets." Let us calculate how occasionally the "very occasionally" will be that a sonnet of Shakespeare will appear.

In *William Shakespeare: The Complete Works*, all Shakespeare's known sonnets are listed. All are quite similar in length. Sonnet number 18 in the list has the well-known opening line: "Shall I compare you to a summer's day?" The sonnet continues through the usual 14 lines, ending as:

> So long as men can breathe or eyes can see,
> So long lives this, and this gives life to thee.

There are 488 letters in the sonnet. Neglecting the spacing between the words, the chance of randomly typing the 488 letters to produce this one sonnet is one in 26^{488}. Using the more familiar 10-base decimal system, the chance is one in 10^{690}. The number 10^{690} is a one followed by 690 zeros! The immense scale of this number is hinted at when one considers that since the Big Bang, 15 billion years ago, there have been only 10^{18} seconds.

To write by random one of Shakespeare's sonnets would take all the monkeys plus every other animal on Earth typing away on typewriters made from all the iron in the universe over a period of time that exceeds all the time since the Big Bang and still the probability of a sonnet appearing would be vanishingly small. At one random try per second, with even a simple sentence having only 16 letters, it would take 2 million billion years (the universe has existed for about 15 billion years) to exhaust all possible combinations.

Chance cannot have been the agent that formed the similarities in such complex things as the nearly identical proteins found in bacteria, wheat, and humans. Somehow we are all related. Furthermore, if life is limited to only a few basic variations in function, one of which is the DNA and RNA genetic system of all Earth-based life, then chance cannot have formed life. There just was not enough time for this to have occurred by chance.

Glossary

■■■■■■■■■■■■■■■■■■■■■■

Absolute zero. The temperature at which the thermal motions within a substance cease. It is the lowest temperature possible. At this temperature a substance no longer possesses any heat. This temperature equals $-273.1°C$, $-459.6°F$, and $0°K$.

Antimatter. Particles of matter with the same mass and spin, but with the opposite charge and other particle characteristics, of the corresponding antiparticle. Electrons and antielectrons, protons and antiprotons are examples of antimatter. When a particle and its antiparticle collide, their mass converts to photons containing a total energy equivalent to their combined mass.

Big Bang. The explosion that theoretically occurred at the beginning of the universe. Since the Big Bang, the universe has been expanding outward from the minuscule volume it occupied at the beginning.

Biosphere. That region on the Earth where there is life. This includes the oceans, the dry land, and the lower regions of the atmosphere.

Black body temperature. The temperature to which a totally black object would have to be heated for that object to emit a particular thermal radiation. For example, radiation with a wavelength of 7.35 centimeters is equivalent to the radiation that would be emitted by a black body heated to a temperature of $3.5°K$.

Black hole. A location with such a high concentration of mass that the gravity produced by this mass does not allow anything, including light, to escape.

Bronze Age. A period starting some 4500 years ago marked by the first appearance of copper and bronze artifacts in archaeological finds. The Bronze Age lasted approximately 1500 years, at which time iron artifacts appeared. This marked the close of the Bronze Age and the start of the Iron Age.

Catalyst. A substance that increases the rate of a reaction, such as the rate of reaction among chemicals. The catalyst is usually not changed by the reaction and so can be used time and again for promoting the same reaction in new batches of the chemicals. Living organisms depend heavily on catalysts speeding the biological reactions within their cells.

Centrifugal force. An apparent force that feels as if an object is being pushed in an outward direction. The occupants of a car feel this when the car goes around a sharp bend. The force is not real. The feeling results from the inertia of the occupants attempting to make them travel in a straight line while the car forces them to travel around the bend.

Cosmology. The study of the origin and characteristics of the universe.

Critical density. The smallest average density of matter in the universe that would cause the expansion of the universe to stop and then be followed by a contraction of the universe. This critical density is required if the universe is in a state of perpetual cycles of expansion and contraction, the oscillating universe theory. At present, the average density of the universe is estimated to be only 10 percent of the critical density. This means that the universe is in a state of unending expansion. The expansion started at the creation of the universe.

Deuterium. A rare form of hydrogen that has in its nucleus one proton and one neutron. The common form of hydrogen has only a proton in its nucleus.

Doppler effect. The change in frequency of any wave of energy (such as sound waves or light waves) caused by motion of the emitter of the energy relative to the receiver of that energy.

Electromagnetic force. One of the fundamental forces of the universe. It is the force that pulls particles of opposite electric charge together. It is about 100 times weaker than the strong nuclear force.

Electron. The lightest of the basic particles of matter. It has a negative charge and orbits the nucleus (which is positively charged) of an atom.

Electron volt. A unit of energy that equals the energy attained by an electron as it travels across a voltage differential of one volt. An object weighing 2 grams and traveling at 1 centimeter per second would have an energy of about a million million electron volts (eV). Clearly, 1 electron volt is a very small unit of energy.

Entropy. A measure of the disorder in any system. An increase in entropy means an increase in the disorder of the system. In almost every action, the entropy increases. It never decreases. This is the second law of thermodynamics.

Frequency of a wave. The rate at which complete wave cycles (crest and trough) pass a given point. Typically this is reported as cycles per second.

Gamma ray. A class of electromagnetic radiation; the identical type of radiation of which visible light is composed, but with wavelengths much shorter than the wavelengths of visible light.

General relativity. The theory that states that gravity and forces that act like gravity, such as acceleration, affect the space-time continuum. For our study, it tells us that changes in gravity affect the rate at which time passes. The theory of general relativity was reported by Albert Einstein in 1916.

Half-period. Also called half-life, it is the time in which half of the nuclei present at the start of the half-period decay. Thus after one half-period, half the nuclei are left. After two half-periods, one-quarter ($\frac{1}{2} \times \frac{1}{2}$) of the nuclei are left, and so forth.

Helium. The second lightest and the second most abundant element in the universe. Its nucleus consists of two protons and two neutrons.

Hydrogen. The lightest and most abundant element in the universe. Its nucleus consists of a single proton. (See also Deuterium).

Isotropic radiation background. A radiation observed coming in equal intensity from all directions of the universe. Arno Penzias and Robert Wilson were awarded the Nobel Prize for their discovery, in 1964, of the isotropic radiation background.

Kelvin temperature scale. A temperature scale in which each division exactly equals the divisions in the centigrade scale, but with the zero value at absolute zero. In the centigrade scale, the zero value is at the melting point of ice (when the pressure is one atmosphere as on the surface of the Earth). In the Kelvin scale, ice melts at 273.1°.

Kinetic energy. The energy associated with the motion of an object.

Light-year. The distance traveled by light in one year. This equals 9.46 million million kilometers.

Mass. The amount of matter in an object, often approximated by its weight. (Weight is actually the force that gravity exerts on an object.) The mass of an object is a measure of its inertia.

Mishnah. A commentary on, or interpretation of, the Torah. It includes discussion of religious and civil laws and customs, some of which are not obvious from a literal reading of the text of the Torah. Its formulation extends over a period from 300 B.C.E. to C.E. 200 when it was committed to writing by Rabbi Yehudah Hanasi. The word *Mishnah* means "repetition, or teaching orally." The Mishnah is included in the Talmud.

Morphology. The branch of biology that deals with the form and structure of plants and animals.

Mu-meson. A particle produced in nature by collisions of photons and atomic nuclei near the top of the Earth's atmosphere. The rest mass of a mu-meson is 210 times that of an electron. It has either a plus or minus charge and decays with a 1.5-microsecond half-period.

Neutrino. A particle with zero mass or almost zero mass and no charge. It has only weak interactions with matter. Large numbers of neutrinos are produced in supernova explosions.

Neutron. A particle that has no charge and is slightly heavier than a proton. It is one of the particles found in the nuclei of all atoms heavier than hydrogen, the lightest atom. When part of an atomic nucleus, the neutron is stable; however, neutrons not bound in a nucleus are radioactive and have an average life of about 15 minutes.

Newtonian mechanics. A description of the physical world in which the rate at which time passes is identical in all systems irrespective of relative motions between those systems.

Nucleosynthesis. The synthesis of atomic nuclei from the more basic particles of protons and neutrons. Nucleosynthesis of helium occurred in the Big Bang. The nucleosynthesis of almost all other elements has taken place, and still takes place, in the cores of stars.

Ontogony. The development of a particular organism over time. The statement "ontogony recapitulates phylogeny" means that during the development of an organism, the embryo and fetus pass through stages in which they have shapes and body parts that were parts of their historical predecessors' bodies. Thus during stages of its growth, the human fetus has gill slits as does a fish, then a 3-chambered heart as does a reptile, and later a tail similar to "lower" mammalian animals.

Orthogenesis. Changes in the form of a living organism that, in each succeeding generation, move the form toward some particular, predetermined (but perhaps unknown to the observers) ultimate form.

Oscillating universe. A version of the Standard Model of the universe that states that the universe has undergone and will undergo an infinite number of cycles of expansion and contraction, each cycle being marked by a Big Bang. The oscillating-universe theory states that there was no beginning and there will be no end to the existence of the universe.

Photon. The quantum or unit of all electromagnetic radiation (of which light is one form). Light may be described as either a wave of energy or as a stream of individual photons.

Proton. A particle with a positive charge. It is found in all atomic nuclei and it is the particle that gives all atomic nuclei their positive charge.

Radioactivity. The breaking of an atomic nucleus into several smaller nuclei. Energy in the form of electromagnetic radiation is emitted as the disintegration occurs.

Red shift. The Doppler effect that causes the spectrum of light emitted by a source moving away from an observer to appear more red than it would have been had there been no relative motion between the emitter and the observer. A red shift in the light of a star is an indication that the star is moving away from the Earth.

Rest energy. The energy that would be released by converting the mass of a particle at rest into its energy form in accord with Einstein's formula $E = mc^2$.

Rest mass. The mass of a particle at rest.

Space-time reference frame. A way of viewing the world in which the four dimensions, length, width, height, and time, are considered simultaneously. As such, a space-time reference frame is four dimensional and not three dimensional, as we are accustomed to when describing space independently of time.

Special relativity. The description of physical systems in which relative motion between those systems affects the perception of size, mass, and the rate at which time passes. The theory of special relativity was published by Albert Einstein in 1905.

Speed of light. Symbolized as c, it equals 299,792,500 meters per second in a vacuum. As particles that have a rest mass approach the speed of light, their mass or weight approaches infinity. Because of this, such particles can never reach the speed of light. At sea level, in air, the speed of light is about 1 million times faster than the speed of sound.

Standard Model of the universe. A description of the origin and development of the universe widely accepted as being correct, hence the name "Standard" Model. The Standard Model states that the existence of space, matter, and the passage of time started simultaneously at the instant of the Big Bang. Since the Big Bang, the universe has been in a state of expansion. The Standard Model does not state whether or not this expansion will continue eternally.

String Theory. A current *theoretical* description of the universe that uses ten dimensions to describe the matter (such as protons and electrons) and the forces (such as gravity and electromagnetism) of the universe. In the String Theory, four of the dimensions are perceivable in everyday life as length, width, height, and time. The remaining six dimensions exist but are of such minuscule size that they cannot be measured.

Strong nuclear force. The strongest of the four fundamental forces of nature. It acts at ranges of 10^{-15} meters and binds the particles in the nuclei of atoms.

Supernova. The explosion of a star. Usually massive amounts of energy are released during the first days of the explosion. A supernova observed in 1987 was so powerful that although the star was 170,000 light-years distant from the Earth, the light of the supernova was visible to the naked eye.

Talmud. Also known as the "Oral Law." A commentary on the Torah that includes the Mishnah and the Gemara. The concise style in which the Mishnah was written made understanding it difficult. The Gemara was added to explain the concepts presented in the Mishnah. Together the Mishnah and the Gemara comprise the Talmud. The text of the Talmud was fixed in about C.E. 500. The word *Talmud* means "study."

Teleology. The situation of being directed toward a given end by some purpose, known or unknown. In nature, teleology might be assumed in the flow of life from primitive forms to higher forms.

Thermodynamics. The portion of physics that deals with actions or relations among heat and matter.

Torah. The Five Books of Moses: Genesis, Exodus, Leviticus, Numbers, Deuteronomy. The Torah is believed to have been revealed to Moses on Mt. Sinai approximately 3400 years ago, two months after the Israelites left Egypt in the Exodus. The word *Torah* means "a showing of the way."

Ultraviolet radiation. Symbolized as uv, it is a form of electromagnetic radiation with wavelengths just shorter than those visible by the eye. In high doses, such as reach the outer atmosphere of the Earth, uv radiation can be lethal.

Wavelength. The distance between the same parts of two successive waves, such as the distance between crests or between troughs. The wavelength of red light is approximately 0.7 millionths of a meter.

Weak nuclear force. One of the fundamental forces of nature. It is 100,000 times weaker than the strong nuclear force. It causes the decay of atomic nuclei and the decay of free neutrons.

Bibliography

■■■■■■■■■■■■■■■■■■■■■

Biblical References

(References in this section of the Bibliography are listed according to the date of the original source. References in the remaining sections are listed alphabetically.)

The Holy Scriptures, Jerusalem: Koren Publishers of Jerusalem, 1969 (Hebrew and English).

Rosenbaum, M., and A. Silbermann, *Pentateuch with Onkelos's Translation (into Aramaic) and Rashi's Commentary*, Jerusalem: Silbermann Family Publishers, 1973 (Hebrew and English).

The Babylonian Talmud, Jerusalem: Eshkol and J. Weinfeld Ltd., 1978 (Aramaic).

Maimonides, *The Guide for the Perplexed*, trans. M. Friedlander, London: George Routledge & Sons, 1928 (English).

Ibid., trans. S. Pines, Chicago: University of Chicago Press, 1974 (English).

Nahmanides, *Commentary on the Torah*, ed. C. Chavel, Jerusalem: The Rav Kook Institute, 1958 (Hebrew).

Ibid., trans. and ed. C. Chavel, New York: Shilo, 1971 (English).

Cosmology and Physics

Atkins, K., *Physics*, New York: John Wiley, 1970.

Badash, L., "The age-of-the-earth debate," *Scientific American* 261 (August 1989): 90–96.

Bloxham, J., and D. Gubbins, "The evolution of the earth's magnetic field," *Scientific American* 261 (December 1989): 68–75.

Bondi, H., *Cosmology*, Cambridge, England: Cambridge University Press, 1960.

Cohen, I. B., *The Birth of a New Physics*, New York: Norton, 1985.

Cohen, M., and I. Drabkin, *A Source Book in Greek Science*, Cambridge, MA: Harvard University Press, 1958.

Crease, R., and C. Mann, *The Second Creation*, New York: Macmillan, 1986.

Dickson, F., *The Bowl of Night*, Cambridge, MA: MIT Press, 1968.

Einstein, A., *The Principle of Relativity*, London: Methuen, 1923. Reprinted by Dover, New York, 1952. (This includes Einstein's 1917 paper, "Cosmological consideration on the general theory of relativity.")

———, *Relativity: The Special and General Theories*, New York: Crown, 1961.

French, A. P., ed., *Einstein, A Centenary Volume*, Cambridge, MA: Harvard University Press, 1979.

Gibbons, A., "Chaos and the real world," *Technology Review* 91 (1988): 10–11.

Gleick, J., *Chaos: Making a New Science*, New York: Viking, 1987.

Gore, R., "The once and future universe," *National Geographic* 163 (1983): 704–49.

———, "Our restless planet: earth," *National Geographic* 168 (1985): 142–81.

Guth, A., and P. Steinhardt, "The inflationary universe," *Scientific American* 250 (May 1984): 116–28.

Hafele, J., and R. Keating, "Around-the-world atomic clocks: predicted relativistic time gains," *Science* 177 (1972): 166–68.

———, "Around-the-world atomic clocks: observed relativistic time gains," *Science* 177 (1972): 168–70.

Hawking, S., *A Brief History of Time*, New York: Bantam, 1988.

Hegstrom, R., and D. Kondepudi, "The handedness of the universe," *Scientific American* 262 (January 1990): 108–15.

Hermann, A., *The Genesis of the Quantum Theory (1899–1913)*, Cambridge, MA: MIT Press, 1971.

Kirshner, R., "Supernova—death of a star," *National Geographic* 173 (1988): 618–47.

Lewis, G., and M. Randall, *Thermodynamics*, New York: McGraw-Hill, 1961.

Newton, I., *Mathematical Principles of Natural Philosophy*, Berkeley: University of California Press, 1947.

Pagels, H. R., *Perfect Symmetry*, New York: Simon and Schuster, 1985.

Rothman T., "Cosmic lithium suggests the universe is open," *Scientific American* 261 (August 1989): 16.

Schilpp, P., ed., *Albert Einstein: Philosopher-Scientist*, 2 vols., Peru, IL: Open Court, 1970.

Schwartz, B., "Muddy evidence," *Scientific American* 260 (June 1989): 13.

Shankland, R., "The Michelson-Morley experiment," *American Journal of Physics* 32 (1964): 16–20.

Taylor, E. F., and J. A. Wheeler, *Spacetime Physics*, San Francisco: W. H. Freeman, 1966.

Velikovsky, I., *Worlds in Collision*, New York: Delta, 1965 (original 1950).

Von Arx, Wm., *An Introduction to Physical Oceanography*, Reading, MA: Addison Wesley, 1962.

Weinberg, S., *The First Three Minutes*, New York: Basic Books, 1977.

Whitrow, G., *The Structure and Evolution of the Universe*, New York: Harper Torchbooks, 1959.

Woosley, S., and M. Phillips, "Supernova 1987A!" *Science* 240 (1988): 750–59.

Woosley, S., and T. Weaver, "The great supernova of 1987," *Scientific American* 261 (August 1989): 32–40.

Life

Alvarez, L., et al., "Extraterrestrial cause for the Cretaceous-Tertiary extinction," *Science* 208 (1980): 1095–108.

Ayala, F. J., "The mechanisms of evolution," *Scientific American* 239 (September 1978): 56–69.

Barghoorn, E., "The oldest fossils," *Scientific American* 224 (May 1971): 30–42.

Benditt, J., "Grave doubts: the Neanderthals may have not buried their dead after all," *Scientific American* 260 (June 1989): 16.

Bürgin, T., et al., "The fossils of Monte San Giorgio," *Scientific American* 260 (June 1989): 50–57.

Churchland, P. M., and P. S. Churchland, "Could a machine think?" *Scientific American* 262 (January 1990): 32–37.

Coles, J., "The world's oldest road," *Scientific American* 261 (November 1989): 100–106.

Crick, F., and L. E. Orgel, "Directed panspermia," *Icarus* 19 (1973): 341.

Darwin, C., *On the Origin of Species*, facsimile edition, Cambridge, MA: Harvard University Press, 1964.

Dobzhansky, T., et al., *Evolution*, San Francisco: W. H. Freeman, 1978.

Dubos, R., *Celebrations of Life*, New York: McGraw-Hill, 1981.

Eldredge, N., and S. J. Gould, "Punctuated equilibria: an alternative to phylotic gradualism," in *Models in Paleobiology*, ed. T. Schopf, San Francisco: Freeman, Cooper, 1972, pp. 82–115.

Fagan, B., *Prehistoric Times*, San Francisco: W. H. Freeman, 1983.

Gore, R., "Extinctions: what caused the earth's great dyings?" *National Geographic* 175 (1989): 662–99.

Gould, S. J., *Ever Since Darwin*, New York: Norton, 1977.

———, *Hen's Teeth and Horse's Toes*, New York: Norton, 1983.

———, *The Panda's Thumb*, New York: Norton, 1980.

———, and N. Eldredge, "Punctuated equilibria: the tempo and mode of evolution reconsidered," *Paleobiology* 3 (1977): 115–118.

Gregor, B., "Carbon and the atmospheric oxygen," *Science* 174 (1971): 316–17.

Hoyle, F., and C. Wickramasinghe, *Lifecloud*, New York: Harper & Row, 1978.

———, *The Origin of Life*, Cardiff, Wales: University College of Cardiff Press, 1980.

Jeffrey, D., "Fossils: annals of life written in rock," *National Geographic* 168 (1985): 182–91.

Kent, G., *Comparative Anatomy of the Vertebrates*, St. Louis: C. V. Mosby, 1983.

Lampert, D., and M. Branwell, eds., *The Brain*, New York: Putnam, 1982.

Leakey, L., and H. Van Lawick, "Adventures in the search for man," *National Geographic* 123 (1963): 132–52.

Leakey, M., "Tanzania's Stone Age art," *National Geographic* 164 (1983): 84–99.

Leakey, R., and A. Walker, "Homo erectus unearthed," *National Geographic* 168 (1985): 624–629.

Marshack, A., "An Ice Age ancestor?" *National Geographic* 174 (1988): 478–81.

Mehringer, P., Jr., "Weapons of ancient americans," *National Geographic* 174 (1988): 500–503.

Miller, G. T., *Energetics, Kinetics and Life—An Ecological Approach*, Belmont, CA: Wadsworth, 1971.

Miller, S. L., "A production of amino acids under possible primitive Earth conditions," *Science* 117 (1953): 528–529.

Morowitz, H., *Energy Flow in Biology*, New York: Academic Press, 1968.

Oparin, A. I., *The Origin of Life*, New York: Macmillan, 1938.

Putnam, J., "The search for modern humans," *National Geographic* 174 (1988): 438–77.

Raup, D., and J. Sepkosi, Jr., "Mass extinctions in the marine fossil record," *Science* 215 (1982): 1501–1503.

Renfrew, C., "The origins of Indo-European languages," *Scientific American* 261 (October 1989): 106–14.

Rennie, J., "In the beginning: evidence grows that RNA was the first self-made molecule," *Scientific American* 261 (September 1989): 28–32.

Rensbergher, B., "Recent studies spark revolution in the interpretation of evolution," *The New York Times* (November 4, 1980): C 3.

Ridley, M. "Evolution and the gaps in the fossil record," *Nature* 286 (1980): 444–445.

Rigaud, J. P., "Art treasures from the Ice Age: Lascaux Cave," *National Geographic* 174 (1988): 482–99.

Schopf, J. W., "The evolution of the earliest cells," *Scientific American* 239 (September 1978): 110–112.

Stanley, S., and C. Harper, Jr., "Stability of species in geologic time, *Science* 192 (1976): 267.

Taylor, G. R., *The Great Evolution Mystery*, New York: Harper & Row, 1983.

Thomas, D. W., ed., *Archaeology and Old Testament Study*, Oxford: Clarendon, 1967.

Thomas, L., *The Incredible Machine*, Washington, D.C.: National Geographic Society, 1986.

Wald, G., "The origin of life," *Scientific American* 191 (August 1954): 44–52.

Weaver, K., "The search for our ancestors," *National Geographic* 168 (1985): 559–623.

Wells, S., and G. Taylor, eds., *William Shakespeare: The Complete Works,* Oxford: Clarendon, 1986.

White, R., "Visual thinking in the Ice Age," *Scientific American* 261 (July 1989): 92–97.

Wilson, E., "Threats to biodiversity," *Scientific American* 261 (September 1989): 108–16.

Index

·····················

Page references for Figures, Tables, and Illustrations are italicized

199